Rational Welding Design

Rational Welding Design

Second Edition

T. G. F. GRAY
Department of Mechanics of Materials,
University of Strathclyde, Glasgow

J. SPENCE
Department of Mechanics of Materials,
University of Strathclyde, Glasgow

Butterworths
London Boston Durban Singapore
Sydney Toronto Wellington

First published in 1975 by Newnes-Butterworths
Second edition by Butterworths, 1982

© Butterworth & Co. (Publishers) Ltd, 1975, 1982

British Library Cataloguing in Publication Data

Gray, T. G. F.
   Rational welding design.   2nd ed.
   1. Welding
   I. Title   II. Spence, J.
   671.5′2   TS227.2

   ISBN 0-408-01200-5

Filmset by Mid-County Press, London SW15
Printed by Page Bros. Ltd., Norwich, Norfolk

Preface to the First Edition

We readily admit that the title of this book is presumptuous—not to say cheeky, but how else could we persuade you to open it? Nevertheless, it is too often true in welding design that irrational rules are substituted for logical thought. Sometimes when a cast component is superseded by a welded fabrication, features of the original casting such as gussets and fillets are quite unreasonably translated to the fabrication. As another example, the concept of 'joint efficiency' which was developed in the context of riveted construction, still lurks in the background of welded structural design. The reader will no doubt be able to add examples from his own experience.

It can be assumed that rational design will in the first place require a thorough understanding of the relevant technologies. Unfortunately, the field of interest is wide enough to be daunting, and if designers occasionally throw in the towel and retreat to simple but often irrelevant rules, it is hardly surprising. Someone who has had a formal training in engineering or metallurgy will find that welding technology stretches his fundamental knowledge to its limits and beyond. To make matters worse, the variety of topics concerned have a subtle interdependence which forces the specialists in each field to come out of their compartments and try to understand each other's problems.

Understanding welding processes and metallurgy presents a particular difficulty. Mechanical engineers traditionally learn machine shop practice at first hand, and as a consequence machined component designs usually display confidence and a sureness of touch. Opportunities to absorb welding practice in the same way are more difficult to obtain, however, leading to a certain vagueness of specification, exemplified in the delightful drawing office instruction 'Weld Here'. (This exhortation has been described as an invitation to a disaster.) In Part I of the book therefore we hope to encourage the designer to build up his knowledge of processes and metallurgy. If the experts in those fields detect some cutting of corners or avoidance of popular controversies, we plead for indulgence, on the grounds that a short clear exposition is more likely to be encouraging.

In any case, welding engineers and metallurgists will often be available to help the designer towards a sound manufacturing proposition. On the other hand, the question of service performance and strength in Part II is very much the engineers' responsibility. Until recently, the central problem of fracture was impossible to treat explicitly by the application of mechanics principles. Hence it was very easy to ask the designer embarrassing questions concerning the possible failure of his structure. What does the 'safety factor' cover? Are welds weaker than parent metal? Should the yield strength of the weld metal be more or less than the parent metal?—and so on. These questions are still troublesome but the timely development of subjects such as fracture mechanics and the increase in understanding of fatigue failure, should put the engineer in a much stronger position to answer them. These subjects are difficult and are not usually treated to a sufficient level in undergraduate courses: we hope therefore that Chapters 5 and 6 will help the designer to break into this area of knowledge.

The uncertainty which engineers may feel about metallurgy has an equivalent among metallurgists who worry about the realism of the mechanical models which engineers employ. A deliberate attempt has been made in Chapters 3 and 4 to make this area more accessible for metallurgists, NDT experts and others who have an interest in the fate of welded structures.

We should feel rewarded therefore if this book helps the reader to develop in his own field, and enables him to communicate better with other specialists.

T.G.F.G.

Preface to the Second Edition

Attitudes to welding design have changed substantially for the better since the preface to the first edition was written seven years ago. Programmes on television have possibly done something to place welding in its proper context as a high-grade technology at the heart of many exciting developments in modern engineering, and there are now many large and visible examples of complex welded structures in the offshore industry to reinforce the arguments.

The rather casual approach to the specification of welding matters which used to be common in many areas of industry has fortunately diminished—the implications of the Health and Safety at Work Act in the UK and a generally sharpened awareness of the hazards associated with load-carrying structures, pressure containments and chemical plant have done much to encourage a more thoughtful approach. It does not therefore seem appropriate to maintain the mildly provocative stance which characterised the earlier preface. Attitudes to welding design are much healthier than they were! However the central argument of the book, as expressed in the title, remains valid and undiminished. Welding design is most effectively approached from as wide an understanding as possible of the various interrelated technologies. It is decidedly not a matter for dull, uninformed rules.

In considering what changes might be needed for the second edition, it seemed to us that most attention should be given to the topics of fracture mechanics and fatigue research which have developed significantly in the last 10 years. Many engineers had not heard of fracture mechanics in the early 1970s, or else they considered it to be a specialist subject of limited application to general engineering. Such an attitude is no longer tenable. The power industry was one of the first to discover that even simple fracture mechanics treatments could be extremely cost-effective in terms of reducing plant outages and increasing real safety. Many other industries have followed suit and fracture assessments are now a common requirement in design procedures. The volume of data on fatigue has also increased greatly over the years and there are now many more calls on such information in relation to the design of offshore structures, chemical plant, cranes and bridges, to name but a few.

We have therefore rewritten and extended Chapters 5 and 6 to include more information, but as in the first edition, our aim has been to be brief in the interest of clarity. Nevertheless, these chapters will in themselves provide a base for practical problem-solving and the approaches are consistent with other treatments now appearing in handbooks and standards.

In conclusion it is clear that the challenge of rational welding design is greater than ever as the variety and scale of welded artefacts increases annually. We have been uplifted over the years by letters from readers who have been encouraged to believe that welding is less mysterious tham they had previously imagined and who have actually been able to use the text to formulate answers to real problems in numerical terms. We hope that the second edition will continue to foster such an approach.

T.G.F.G.
J.S.

Acknowledgements

Most of the material for this book was developed while teaching various postgraduate and post-experience courses in Welding Technology at The University of Strathclyde. We are grateful to the Companies which have supported these courses with enthusiasm, and to the many students who contributed a wide practical experience to the discussions.

The present authors are glad to acknowledge the contributions of Dr Tom North to the first edition, substantial portions of which have carried over to the second edition although he has moved abroad to an industrial post in the intervening period.

Thanks are also given to Janet Harbidge and Mary Lauder who patiently typed and re-typed the manuscript many times.

The cover photograph was printed by courtesy of British Federal Welder & Machine Company Ltd, Dudley, West Midlands.

Contents

PART I. DESIGN FOR WELDING

Most designers think that welding is related to black magic.... A few do try to adjust their designs to help the welding engineers, but when they look into the subject they find it too confusing and so abandon hope.

R. BOLER
(Nuclear Power Group)

Chapter 1

Welding processes and their influence on design

1.1 Basic requirements of a welding process

A weld can be simply described as a region of bonding between adjacent solids. Ideally the mechanical, metallurgical and chemical properties of the weld should be identical to those of the solids which are joined, but this hardly ever happens. A bewildering variety of welding processes and techniques exist, having different applicabilities and characteristics. Nevertheless, they have a common purpose, and should resemble each other at some basic level. The important characteristics of a given process and its relationship to others can best be understood by considering the primary requirements of a bonding operation.

Spontaneous metal-to-metal bonding will occur if perfectly clean and smooth metal surfaces are brought into intimate contact; that is, if the atoms of one surface are brought to within atomic distances of the other. In practice, the surfaces of metals in the welding shop are covered with an adsorbed oxide layer and are rough. Thus intimate contact cannot be made. A welding process, therefore, has only to remove the oxide layers, and bring the surfaces into close contact, for a bond to be made.

The oxide film can be broken down and removed in a number of ways. Mechanically, or by chemical fluxing, or by sputtering in an electric arc, or by dissociation and evaporation in a combined high-vacuum/high-temperature environment. In some processes the oxide film is disrupted mechanically, for example by compressing the joint surfaces as in cold-pressure welding, or by rubbing the surfaces together as in friction welding, or by deforming the joint interface at high rates as occurs in explosive welding. The dissociation and evaporation technique is used in diffusion bonding. In fusion welds, the oxide film floats off on the molten weld pool and can be disrupted by a combination of arc action and/or fluxing.

Intimate contact in the solid-state processes is produced by mutual plastic deformation. Many of the rubbing and deforming processes, described later, generate heat. This reduces the yield strength and allows easy plastic flow. In the fusion processes the adjacent surfaces are melted and linked by a molten bridge which solidifies to form a continuous structure. Extra molten metal may be added to the joint.

The direct addition or indirect evolution of heat during welding is the key to many processes, but it is at the root of many of the welding engineer's problems. The mechanical and metallurgical properties of the joint material very often suffer. In many respects, the melted zone in a fusion weld resembles a miniature casting and the adjacent unmelted parent material in what is called the 'heat-affected-zone' (HAZ) is also subjected to a rapid heat treatment cycle which is hardly ever beneficial. Solid-state processes have a distinct advantage in this respect, as the metal should not reach its melting point in a properly made joint. In fact, a few of the solid-state processes are carried out cold.

The mechanical properties of welds are usually impaired if oxygen, hydrogen or nitrogen are absorbed during welding. Once again, high temperatures accentuate the problem, as gases are more readily absorbed. It is also important to prevent the large-scale regeneration of oxide films which might be trapped in the metallurgical structure and weaken it. Another indirect but basic requirement for a welding process, therefore, is that air should be excluded from the joint region during welding. Many different techniques are used—for example welding can be carried out in a vacuum, or the molten zone can be covered by a homogeneous flow of protective gas which may be inert or reducing in character. (Shielding gas composition also influences the behaviour of electric arcs and the character of metal transfer from filler wires if they are used, and therefore, the choice of gas is not completely arbitrary.) Frequently the protective cover is provided by a molten slag bath which floats on the weld pool surface. In certain cases, for example resistance spot welding, air is excluded by the close fit of the parts to be joined.

It is not appreciated that welding, which often appears to be a rather dirty process, requires clean surfaces. The exact needs vary from one process to another, but if any of the common workshop contaminants such as oil, water, heavy rust or dirt, are drawn into the joint, pores or other unwanted defects can easily arise.

1.2 Important features of common welding processes

Welding processes can be categorised into fusion welding processes or solid-state bonding processes. In the case of fusion welding, the aim is to produce a region of molten metal between the workpieces

5

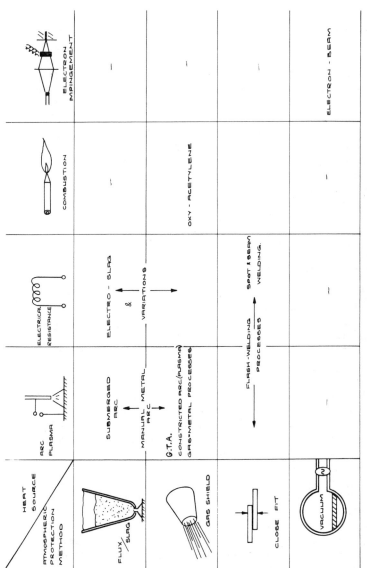

Figure 1.1 Fusion welding processes (solidifying liquid bridge)

and bonding occurs when this molten region solidifies. In 'auto-genous' fusion welds no filler metal is added to the molten pool during the welding operation. When filler metal is added, the chemical composition of the filler metal has the major influence on the final weld composition and consequently its metallurgical and mechanical properties. This flexibility in control of the joint prop-erties is a great advantage of fusion welding since the weld com-position can be varied to suit the application. Fusion welding processes include the electric-arc processes, oxy-acetylene welding, electron-beam and laser welding.

Bonding in solid-state welding processes is achieved with little or no melting as the clean metal surfaces are brought into intimate contact so that metal-to-metal bonds can be formed. Solid-state processes include hot or cold pressure welding, ultrasonic welding, friction welding and explosive welding. The common features of the more important welding processes are illustrated in *Figure 1.1.*

1.2.1 Fusion welding processes

The electric arc as a heat source

The electric arc was first developed for welding in Russia at the end of the nineteenth century. The arc converts electrical energy into thermal energy by a counter-current transfer of electrons to the anode region and positive ions to the cathode region. (Positive ions are atoms or molecules which have lost an electron. They have a net positive charge and consequently move towards the cathode region under the application of an electric field.) The ionisation potential is the energy which must be supplied to produce an ionised atom or molecule. Ionisation occurs when atoms, molecules, and high-velocity electrons which have been emitted from the cathode, collide.

Electrons can be emitted at the cathode region by thermionic emission and the rate of emission of electrons is determined by the absolute temperature and by the work function of the cathode material. The work function of a metal represents the amount of energy (in electron volts) which must be supplied to stimulate emission of electrons from the metal. Thermionic emission of electrons occurs readily in carbon and in tungsten arcs, but is not the main effect in other metals which would reach their boiling points before they could emit electrons thermionically. The mechanism controlling electron emission in the case of these other metals is a matter of conjecture at the present time and a careful examination of recent theories is beyond the scope of this book. The arc is usually started by touching and withdrawing the electrodes. Exceptions will be noted later.

An advantage of the electric arc as a heat source in fusion welding is that the energy conversion occurs with the workpiece as one of the electrodes. In the case of oxy-acetylene welding on the other hand, the heat source is external to the workpiece; the heat evolved from exothermic chemical reactions has to transfer across the gap between the torch and the workpiece. As a consequence, the heat transfer efficiency in electric arc processes is much higher than that in oxy-acetylene welding and the workpiece distortion is much less.

Positional welding

The liquid bridge has, of course, to be held in place satisfactorily until it solidifies. This may be simple enough when welding on a flat horizontal surface where the liquid is supported from below. If a joint has to be made on a vertical surface, or on a horizontal surface overhead, the force of gravity may make welding difficult, particularly if the weld pool is large. The problem will be obvious enough to readers who have tried to apply paint to walls and ceilings. Very often welds have to be made in such positions, however, and it is, therefore, important to be aware of the 'positional' capability of a certain welding process. The commonly used descriptions in order of difficulty are shown in *Figure 1.2*. More detailed information on positional capability is given in *Table 1.1*.

Manual-metal-arc welding

Early work on arc welding was carried out using carbon and tungsten electrodes which were not consumed during the welding operation. Solid bare-wire electrodes were used in the earliest attempts to produce consumable electrodes which could melt off during the welding operation and supply filler metal. However, the replacement of carbon and tungsten electrodes by bare-wire electrodes introduced major problems since arc initiation was difficult, the arc was unstable and the resulting welds contained high levels of oxygen and nitrogen. To reduce these difficulties, bare-wire electrodes were coated with a mixture of refractory oxides, alkali silicates and deoxidants. Electrodes with such refractory coatings are referred to as 'manual-metal-arc' electrodes and *Figure 1.3* illustrates the typical arrangement for welding. The functions of the electrode coating are as follows:

1 Improvement of the arc physical properties—the coating improves arc initiation and arc stability

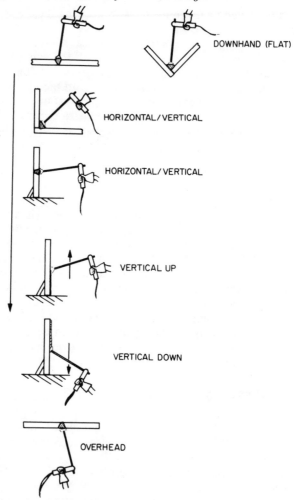

Figure 1.2 Positional welding

2 Exclusion of air from the region of the joint during welding—the coating generates a gas shield or a slag layer on the weld pool surface

3 Improvement of the weld metallurgical properties—the coating allows transfer of alloying elements across the arc and favours formation of high-quality weld metal which has low-oxygen, low-nitrogen and (for some electrodes) low-hydrogen contents.

Arc initiation and subsequent arc stability are enhanced when

Table 1.1 CHARACTERISTICS OF WELDING PROCESSES

Process	Manual-metal-arc			Fusarc
	Class II, III rutile electrodes	Class VI basic electrodes	High-yield electrodes (iron powder)	
Arc shielding method	Electrode coating produces gas or slag shielding			
Power supply	D.C. or A.C.	D.C. or A.C.	D.C. or A.C.	D.C.
Range of currents used (A)	Up to 500	Up to 500	Up to 500	200–1200
Deposition rates (kg/h)	Up to 5	Up to 5	Up to 6.5	1.5–10
Thickness weldable	3 mm to any thickness			10 mm to any thickness
Welding position	All-positional	All-positional	Downhand and horizontal–vertical	
Operating considerations	Labour intensive Little capital investment	Flexible in operation Little maintenance required		Long weld runs recommended
Possible defects	Porosity	Slag inclusions	Cracking in the weld and HAZ region	

Table 1.1 (continued)

Process	Gas-metal-arc			No gas process (deoxidation elements added to wire)	Gas-tungsten-arc
	Spray-transfer type	Dip transfer	Pulsed arc welding		
Arc shielding method	CO_2 gas or argon	CO_2 gas	Argon or argon + CO_2	No gas shield	Argon or helium
Power supply	D.C.	D.C.	D.C.	D.C. or A.C.	D.C. (A.C. for aluminium)
Range of currents used (A)	200–900	50–200	50–300	200–600	40–250
Deposition rates (kg/h)	3–7.5	0.5–5	0.5–7	3–13.5	Slow completion rate
Thickness weldable	5 mm to any thickness	2–15 mm	1.5–40 mm	5 mm to any thickness	0.5–4 mm
Welding position	Downhand and horizontal—vertical	All-positional	All-positional	All positions except overhead	All-positional
Operating considerations	Wind and draughts may cause porosity formation in welds. Good quality labour required with adequate welder training. Equipment requires maintenance. Wire must be clean if low-hydrogen capability is to be realised.			No problems caused by winds and draughts	As for spray transfer etc.
Possible defects	Porosity and cracking in narrow welds	Lack of fusion defects (cold shuts). Porosity	Porosity	Slag inclusions	Porosity, tungsten inclusions, solidification cracking in some autogenous applications

Table 1.1 (continued)

Process	Submerged arc		Electroslag	Electrogas
	Single wire	3-arcs in tandem		
Arc shielding method	Flux covering on weld pool		Flux covering on weld pool	CO_2 gas shield
Power supply	D.C. or A.C.	D.C. or A.C.	D.C. or A.C.	D.C.
Range of currents used (A)	200–1200	200–1200	300–700	400–650
Deposition rates (kg/h)	1.5–15	up to 35	5–22.5 (per electrode wire)	5–22.5 (per electrode wire)
Thickness weldable	3 mm to any thickness	15–25 mm in a single pass	12 mm to any thickness	8 mm–75 mm
Welding position	Downhand and horizontal—vertical		Vertically up	Vertically up
Operating considerations	Long weld runs recommended. If flux-backing systems are used these are expensive		Restarts are difficult if the equipment breaks down during welding	Winds and draughts cause porosity formation
Possible defects	Porosity. Slag inclusions. Weld metal cracking		Centreline cracking in the weld. Under cut. Lack of fusion	Porosity

electrode coatings contain materials which have low ionisation potentials and low work-function values. The chemical compounds of sodium and potassium have low ionisation potentials and for this reason sodium and potassium silicate binders are used during electrode manufacture. The oxides of titanium and magnesium have low work-function values and are normally used in electrode coating formulations.

Protective gases are produced by the decomposition of coating constituents and are generally mixtures of carbon monoxide, hydrogen, carbon dioxide and water vapour. The exact composition

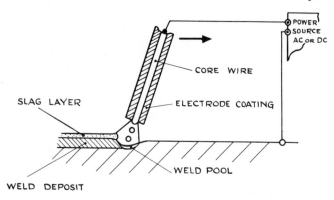

Figure 1.3   The manual-metal-arc welding process

of the shielding gas depends on the formulation of the electrode coating, and gas evolution is favoured by the presence of constituents such as cellulose, rutile and limestone. Slag shielding is provided by the formation of a protective layer of molten oxides on the weld pool surface. This acts as a barrier which prevents contamination of the weld metal. Silica, alumina, rutile, magnesia, fluorspar, lime and manganese oxide are common slag-forming constituents in electrode coatings. After welding, the solidified slag layer has to be removed from the weld surface. It is important that this can be done easily since any slag that is left produces inclusions in subsequent weld runs. Improved deslagging is favoured by the presence of rutile, limestone, fluorspar, manganese oxide and zircon in the electrode coating.

Electrode coatings normally contain additions of deoxidants such as ferro-silicon, ferro-manganese and ferro-titanium. These deoxidants give a further protection for the weld metal against contamination due to breakdown of the gas or shielding systems. The amounts used are carefully controlled, since the addition of too much deoxidant produces weld metal with poor notch toughness.

The electrode coating also provides a means of transferring alloying elements into the weld pool during welding so that any selected weld composition can be produced. Such transfer of powdered alloy additions from the electrode coating is cheap and efficient, and is generally a satisfactory alternative to an alloyed core-wire. This is especially important in cases where the desired weld composition demands an alloy analysis which cannot be drawn into a wire form or is too expensive.

Types of electrode coatings

The chemical compositions and classification society specifications for various types of electrode coating are given in *Table 1.2*. A brief description of the most frequently used electrode classes follows.

Class I cellulosic electrodes. The presence of cellulose in the electrode coating produces large volumes of gas in the arc region. This electrode type relies principally on gas shielding via a mixture of H_2, CO, CO_2 and H_2O. Hydrogen evolved from the electrode coating dissociates in the arc region and then recombines to evolve heat at the weld surface. For this reason, the cellulosic electrodes are very forceful and deep-penetrating. The arc voltages used are in the range 40 to 50 V and the process is characterised by much spatter. Power supplies can be ac or dc and weld metals produced using cellulosic electrodes have hydrogen contents between 30 and 100 ml/100 g of weld metal. This high hydrogen level inhibits the use of cellulosic electrodes for low-alloy constructional steels, because of the risk of hydrogen embrittlement (this phenomenon will be further explained in Chapter 2).

The deep penetration and quick freezing weld pool makes these electrodes especially suitable for vertical-down and overhead work, and therefore they are commonly applied in pipeline welding, where the 'stovepipe' technique is used for circumferential butt welds. Of course, suitable precautions must be taken against hydrogen embrittlement in susceptible materials.

Classes II and III rutile electrodes. Slag formation is the main shielding mechanism in rutile-based electrodes, although there is some gas shielding from hydrogen and carbon monoxide which are present in the arc atmosphere. Rutile electrodes are easy to manipulate, give low spatter and produce welds with well-shaped profiles. Slag formed during the welding operation is easy to remove. The

Table 1.2 ELECTRODE COATING FORMULATIONS OF MANUAL-METAL-ARC ELECTRODES

British specification	European DIN specification	American specification	Electrode designation	Electrode coating formulation
Class 1	DIN Ze	ASTM 6010	Cellulosic	20–60% rutile. 10–50 % cellulose. 15–30% quartz. 0–15% carbonates. 5–10% ferro-manganese.
Class 2	DIN Ti	ASTM 6012	Rutile	40–60% rutile. 15–25% quartz. 0–15% carbonates. 10–12% ferro-manganese. 2–6% organics
Class 3	DIN Ti	ASTM 6013	Rutile	20–40% rutile. 15–25% quartz. 5–25% carbonates. 12–14% ferro-manganese. 0–5% organics
Class 4	DIN Es	ASTM 6020	Acid–Ore	Iron ore–manganese ore. Quartz. Complex silicates. Carbonates. Ferro-manganese
Class 5	DIN Ox		Oxidising	Iron ore. Manganese ore. Quartz. Complex silicates
Class 6	DIN Kb	ASTM 7015	Basic	20–50% calcium carbonate. 20–40% fluorspar. 0–5% quartz. 0–10% rutile. 5–10% ferro-alloys

hydrogen level found in welds deposited using rutile-based electrodes is in the range 15–30 ml/100 g of weld metal. Electrodes of this type are very popular and are used with ac or dc power supplies on many constructional applications.

Class VI basic electrodes. These electrodes are used to produce weld metals with low hydrogen contents, giving a measure of freedom from hydrogen embrittlement in alloy steels. Since coating constituents such as clays, retain their moisture at temperatures in excess of practical electrode baking temperatures, these should be excluded from the coating formulations of basic electrodes. Limestone is a logical choice for basic-coated electrodes because it has favourable arc-stabilising characteristics. It also evolves carbon dioxide which can provide a gas shield. However, a major disadvantage of limestone is its high melting point. This is counteracted by additions of fluorspar (calcium fluoride) which helps to form a lower melting point slag. Hydrogen bearing constituents are not completely eliminated from the formulation of basic electrodes, however. Small quantities of clays are added to ease manufacture and to improve electrode performance. Nevertheless, the hydrogen level in welds produced using properly stored basic-coated electrodes is less than 10 ml/100 g of weld metal.

The gas shield generated by basic coatings is a mixture of carbon monoxide and carbon dioxide. However, the quantities evolved are small, and the arc length must be kept short to avoid air entrapment in the weld. For this reason basic electrodes are more difficult to manipulate than rutile or cellulosic, and require greater welder skill. The slag formed helps to reduce the sulphur and phosphorus content of the weld metal, but it is more difficult to remove after welding.

Basic coatings will absorb moisture if they are left in the open air for any length of time and special care should be taken to keep the electrodes dry. Typical measures include heated stores, sealed electrode containers, small ovens close to the working point, and heated 'quivers' for each welder.

Advantages of the manual-metal-arc process

The arc is visible and welds can be made in all positions (overhead, vertical, etc.) with suitable choice of electrode. The weld metal composition can be varied through the coating formulation.

Metal thicknesses from 3 mm upwards can be welded. Typical deposition rates are shown in *Figure 1.4*.

Mild steels, low-alloy steels, stainless steels, and some nonferrous alloys of aluminium and copper can be welded.

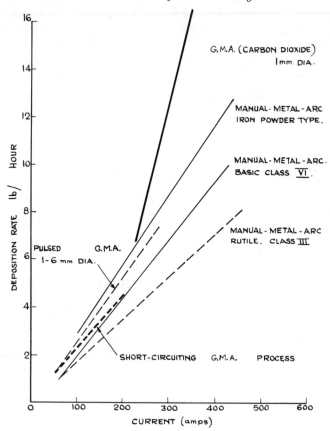

Figure 1.4   A comparison of deposition rates using different electric-arc processes.
(After McMahon, B. P., *Welding and Metal Fabrication*, Jan., Feb., Apr. 1970)

The capital cost of equipment is relatively small, it costs little to maintain, and is robust.

Disadvantages of the manual-metal-arc process

Welds must be deslagged after welding, and if this is not done carefully in multirun joints, there is a risk of slag inclusions in the completed joint.

It is a completely manual technique, and this sets an upper limit on the welding current which can be used of about 500 A. There are some partly mechanised versions of the process, however, in which higher currents and deposition rates can be used. The 'Fusarc' technique is one of the most successful of these methods. In this development, the short straight electrodes are replaced by a continuous coil of coated electrode which is fed to a welding head by drive rolls. Welding current is transmitted to the arc through copper contacts and a fine wire mesh which is wrapped round the coating. Welding currents up to 1000 A can be used, and deposition rates are therefore high.

The submerged-arc process

This process is normally fully mechanised, that is the machine controls the arc length and the movement of the electrode along the joint. A continuous electrode is fed to the welding head from reels

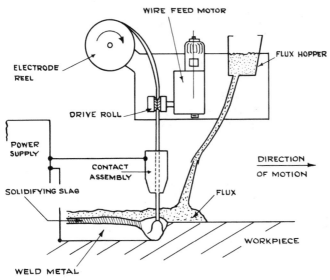

Figure 1.5   The submerged-arc welding process

containing up to 200 kg of wire and electrical power is transmitted to the wire through copper contact tips close to the arcing point. The process acquires its name from the granulated flux blanket which is spread over the joint area, concealing the arc and weld. The flux blanket stabilises the arc and excludes air from the joint. After welding, the flux which has melted and solidified as slag is removed, and excess unmelted flux is recovered for further use. *Figure 1.5* shows a typical arrangement.

The main characteristics of the process are:

It is mechanised, and therefore used to best advantage if the time taken to set up and align the equipment is offset by rapid completion of long weld seams.

The arc is completely invisible, which is more comfortable for the operator, but prevents him from seeing the position and quality of the weld as it is made. The machine must, therefore, be precisely aligned and adjusted before welding.

Accurate weld preparations must be provided as gaps caused by poor fit of the parts cannot be bridged.

Welding currents range from 200–2000 A, and travel speeds of up to 5 m/min are possible.

The process is always applied in the downhand or horizontal–vertical position, because the typically large molten weld and slag pool would be uncontrollable in other positions. Cylindrical components (which for the same reasons must not be too small in diameter) are rotated under a stationary welding head to maintain the downhand position.

Typical weld metals have a low hydrogen content comparable with properly dried class VI electrodes.

A number of variations on the standard electrode arrangement have been developed to increase deposition rates or for cladding applications. Some of these arrangements are shown in *Figure 1.6.* The 'twin-arcs in-series' arrangement minimises penetration into the parent material and is therefore suitable for cladding one material on another where an extensive mix of the two materials is metallurgically undesirable. Plate thicknesses as much as 40 mm have been welded in a single pass by systems incorporating ten simultaneous submerged arcs.

Submerged-arc electrodes are supplied in a number of forms—as bare wire, as 'Fusarc' electrode, as cored wire, or in a strip form. Bare wires, varying from 2.5 to 8 mm diameter, are copper coated to minimise corrosion in store and to improve current transmission. Cored electrodes consist of a tubular metal sheath, which encloses a mixture of powdered flux, deoxidants and alloying elements. Weld metal compositions which would be difficult if not impossible to supply in a bare wire form (due to wire drawing problems) can be obtained, as the troublesome constituents can be mixed in the powdered core. Strip electrodes (typically $60 \times \frac{1}{2}$ mm) are useful for low-penetration cladding.

Submerged-arc fluxes have electrical, physical and metallurgical functions which are similar to manual-metal-arc coatings, and the typical constituents are, therefore, similar. The molten slag should have a suitable surface tension to form the weld bead, and a lower melting point than the weld metal for efficient shielding. Detrimental elements such as sulphur and phosphorus are drawn out by the flux, and alloy additions can be made. For stabilised stainless steels

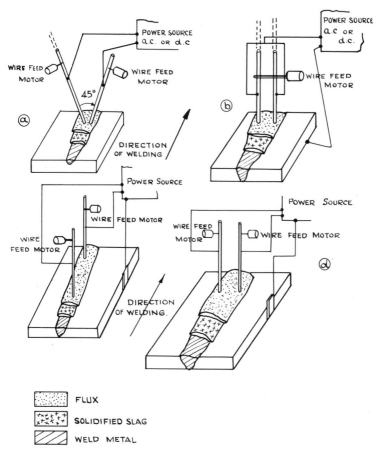

FLUX

SOLIDIFIED SLAG

WELD METAL

Figure 1.6   Multiple electrode systems for submerged-arc welding. (a) Twin arcs in series, (b) twin arcs in parallel, (c) multiple arcs in tandem, (d) multiple arcs in transverse (Knight, D. E., Welding Journal, Apr. 1954)

(see p. 115), the stabilising element niobium is added to the flux. Suppliers will describe fluxes as 'fused' or 'agglomerated'. Fused fluxes are dry-mixed, fused in a crucible at 1200–1300 °C, and quenched in water. They resist moisture pick-up during storage. Unfortunately, alloy additions cannot be made as they would be oxidised at the high fusion temperatures needed. Alloy additions can be made to agglomerated fluxes, in which the constituents are mixed with a sodium or potassium silicate binder, and baked at 400 °C.

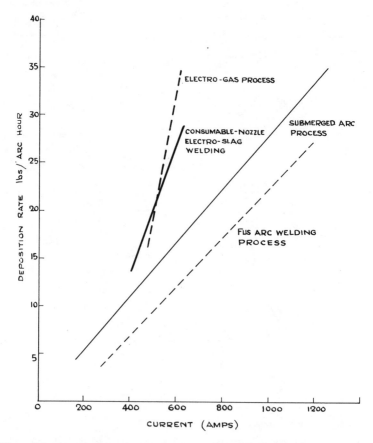

Figure 1.7   A comparison of weld deposition rates using different fusion processes. (After McMahon, B. P., *Welding and Metal Fabrication*, Jan., Feb., Apr. 1970, and Matsuoka, T., *Welding Engineering*, 1966)

Submerged-arc welding is applicable to steels and some non-ferrous metals such as titanium and aluminium. Plate thicknesses greater than 6 mm are suitable and production rates are high as shown in *Figure 1.7*.

The gas tungsten arc (GTA process) (TIG in Britain, WIG in Germany)

In this process, an arc is struck in an inert gas atmosphere, between a pointed tungsten electrode and the workpiece (see *Figure 1.8*). The electrode is *nonconsumable* and, if extra filler metal is required, an electrically neutral or 'cold' wire is fed into the weld pool. No flux is used, and no slag is formed. A flow of argon or helium shields the weld zone and provides the arc atmosphere. The choice of gas is very

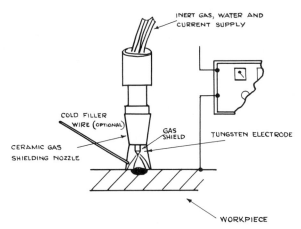

Figure 1.8   The gas tungsten arc process

much decided by cost and availability—in Britain, argon is usual. In the USA, helium is used, and as the arc voltage is consequently higher, greater penetration and speed is obtained.

The common touch-and-withdrawal method for arc starting is *not* used for the GTA process, as the electrode would be contaminated and harmful tungsten inclusions would be trapped in the weld. Instead, a train of high-frequency sparks, or a brief high-voltage surge is used to ionise the gas and start the arc.

Normally, direct current is used, with the electrode negative to minimise electrode heating and to obtain a stable arc shape. However, aluminium cannot be welded in this way due to the presence of a tenacious oxide film on the weld pool surface. When the tungsten electrode is connected as the positive pole this film is disrupted. However, use of a dc current with the electrode positive, overheats the electrode tip and tungsten inclusions can consequently be trapped in the weld. As a compromise, alternating current is usually used for aluminium welding, with a superimposed high-voltage surge or spark to assist reignition of the arc at each polarity reversal. Recent developments include pulsed current supplies to increase control of the heat input, and the use of electrical resistance heated ('hot') filler wires.

Characteristics of GTA

Clean welds of excellent metallurgical quality can be produced, with precise control of penetration and weld shape in all positions, even in thin materials. As shielding is good, reactive metals such as aluminium, magnesium, titanium, niobium and beryllium can be joined. Typical welding currents are low (less than 200 A) and, therefore, weld completion is slow. However, it may be possible to make use of the excellent controllability for difficult stages (for example the root of a butt weld), and use a more productive process for weld completion. Pipe butt welds in nuclear and boiler power plant are usually made in this way.

Essentially, the GTA process (especially in an automatic version) is a precision technique, and demands accurate joint preparation and clean, draught-free workshop conditions.

The plasma-arc process

This process is closely related to the gas tungsten arc process, the significant difference being that the arc plasma is constricted by a nozzle to produce a high-energy plasma stream, in which temperatures over 50 000 K can be reached.

The two basic configurations of plasma torch are illustrated in *Figure 1.9.* In the transferred-arc device, a high-velocity plasma stream passes from the tungsten cathode to the anode workpiece and is constricted by the narrow orifice between the two. In the nontransferred-arc configuration, the torch nozzle forms the anode, and the plasma stream is projected from it at high velocity. Various modifications have been made to the basic nozzle design to give increased control on the plasma stream shape and velocity. These include different orifice patterns and secondary focusing and stabilising gas streams.

The plasma temperature depends on the gas used to form the stream, and gases such as argon (sometimes with 5% hydrogen addition), helium and nitrogen are used. A separate concentric gas flow is provided to shield the molten area.

The very high plasma temperatures and forces can punch through relatively thick material to form a keyhole, the molten metal flowing round the hole edges to solidify behind the arc. Thus a deep narrow weld bead can be produced with consequently lower distortion. Cold filler metal can be added separately as in the GTA process.

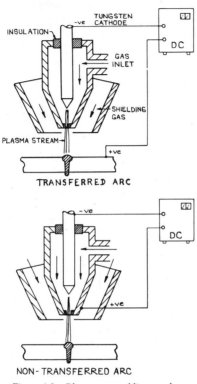

Figure 1.9   Plasma-arc welding torches

The process is sophisticated, and the equipment requires careful maintenance. However, it is very suitable for high-quality welds in difficult materials and can penetrate large thicknesses in a single run. This feature has led to its application in the continuous production of welded pipe.

The gas-metal arc processes

The components of the gas-metal arc process are a continuous filler wire electrode (usually electrically positive) which is fed into an arc between the wire and the workpiece, the whole area being protected by a homogeneous gas flow, (see *Figure 1.10*). (In so-called 'semi-automatic' versions, wire is fed to a manipulated gun, whereas in 'automatic' applications, the welding head is attached to a traversing machine.)

The character of the process depends partly on the gas used. Inert gases such as argon and helium form stable arcs with little weld spatter, and produce high-quality, smooth welds. They are, however, expensive, as the volumes consumed are large. So-called 'active' gases such as carbon dioxide and nitrogen, are much cheaper but

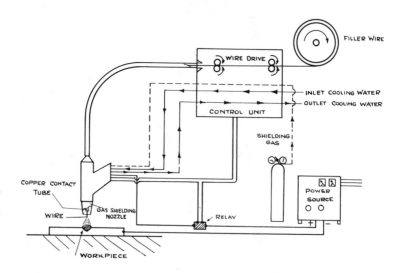

Figure 1.10   The gas-metal arc process

are characterised by increased spatter and less smooth filler wire droplet transfer through the arc. The droplets are partly oxidised in dissociated carbon dioxide arcs, and it is usual to incorporate deoxidants in electrode wires for this application. Mixtures of the various gases are tailored to specific needs, and small proportions of oxygen and chlorine may be added.

Much attention has been given to droplet transfer mechanics and at least three modes have been observed in high-speed photographic films of gas-metal arcs—the spray mode, gravitational transfer, and shortcircuiting ('dip') transfer. The transfer mode obtained in a given case influences the possible application, and depends on arc voltage and current, gas atmosphere composition, and the composition and diameter of the electrode wire. Basic gas-metal welding equipment can, therefore, be used to produce welds of widely differing character, and categorisation is not easy.

The spray transfer mode is associated with inert gases, high currents and small wires (less than 2 mm diameter). Fine droplets are projected across the arc in a stable manner to give well-shaped spatter-free welds (*Figure 1.11a*). At lower currents and/or larger diameters, the droplets are typically larger than the wire diameter, and gravitational transfer occurs.

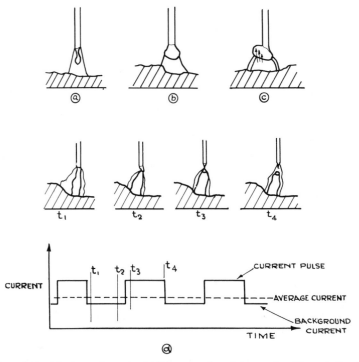

Figure 1.11   Metal transfer modes. (a) Spray-type free-flight transfer; (b) gravitational free-flight transfer; (c) repelled free-flight transfer; (d) transfer during pulsed arc welding

These welds are often rougher and accompanied by spatter. Abrupt mode transitions can occur as shown in *Figure 1.12* which relates to the welding of aluminium with a 1.6 mm diameter electrode in argon. (Whichever mode is used, it is important that the transfer character should be stable within that mode, otherwise inconsistent penetration and shape will result.) In the steel/carbon-dioxide system, droplets experience repulsive forces and transfer is nonaxial as shown in *Figure 1.11c*. The spatter generated is often sufficient to clog the gas shielding nozzle, and frequent cleaning is required.

In the shortcircuiting mode, which occurs at lower currents and voltages (less than 200 A for steel/carbon dioxide), the droplet is not projected across the arc, but grows in size until the arc gap is bridged and an electrical shortcircuit occurs. The rapid current rise

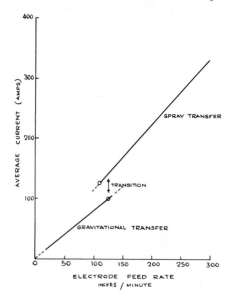

Figure 1.12   Burn-off characteristics for $\frac{1}{16}$ inch diameter pure aluminium in argon. (Needham, J. C., Proc. Symp. on Physics of the Welding Arc, Oct. 1962)

separates the drop, and re-establishes the arc. This is illustrated in *Figure 1.13*, which also shows typical current and voltage cyclic variations. In this low-current régime, the weld pool is small, and the process finds ready application for the all-position welding of sheet steels in car bodies and similar components. There is some risk in applying this low heat input process to steels thicker than 6 mm as lack-of-fusion defects may occur, which are difficult to detect.

Further versatility has been provided by special pulsed power supplies. In this development, artificial current pulses are super-imposed on a low background current, which on its own would produce gravitational transfer. In this way spray transfer can be obtained at average currents, which are below the normal transition range—each droplet transfer coinciding with a current pulse. As the

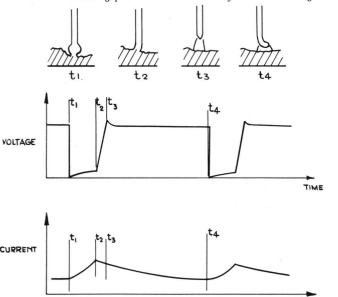

Figure 1.13 Shortcircuiting (dip) transfer in the carbon-dioxide-shielded process

heat input rate is thus reduced, this improvement is especially useful for thin-sheet applications where distortion and excessive melting would otherwise prevent the application of the gas-metal arc process. This variant is illustrated in *Figure 1.11d*.

Characteristics of gas-metal arc processes

These processes require an operational precision which is more exacting than manual-metal-arc or submerged-arc techniques. The equipment used must have an accurate mechanical and electrical performance and must be carefully serviced. Weld bead sizes are usually smaller than submerged-arc deposits, and must, therefore, be more accurately placed. The gas shield integrity, as in the GTA process, is susceptible to strong draughts.

With appropriate choices of shielding gas and transfer mode, the processes can be applied to the widest variety of materials and thicknesses. (It should be noted that carbon-dioxide shielding gas should not be used for steels which rely on carbon levels below 0.15%, for example stainless steels, because the weld metal acquires carbon from the gas. Similarly, high-carbon steels are oxidised by the dissociated CO_2, producing porosity.) Inert gases are expensive but are necessary for reactive materials such as aluminium.

Metallurgically clean, low-hydrogen deposits are typical, without significant slag. Certainly deslagging between weld runs can be largely avoided in multirun applications. On the other hand, alloying elements cannot be added separately as in flux-using processes, and indeed alloy losses in the arc may have to be compensated in the wire analysis.

The electro-slag process

For butt welds which can be presented during fabrication in a vertical configuration, the electro-slag process is extremely attractive. Joints thicker than 15 mm (without an upper limit on thickness) can be welded in one pass, and a simple square-edged joint preparation is used.

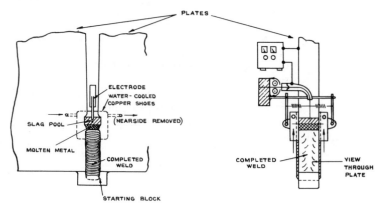

Figure 1.14a   The electro-slag welding process

The essential components of an electro-slag welding machine are illustrated in *Figure 1.14a*. The process greatly resembles a vertical casting operation since the molten weld metal is contained by the parent plates on two sides, and by an opposed pair of cooled copper shoes on the other two. There is no arc. The continuously fed electrode is melted off by resistance heating as it passes through a conductive molten slag bath. This bath also melts the adjacent parent plate edges, and excludes the atmosphere from the molten metal. As welding proceeds the copper shoes and wire-feed machinery are moved slowly up the joint at speeds of the order of 30 mm/min.

Metal deposition rates are about 20 kg/h. For thicker plates, more wires are spread across the joint thickness. The beginning and end of the weld are usually defective on account of lack of fusion and slag inclusions. Starting and finishing blocks which are cut off afterwards are therefore used. Special closing and overlapping techniques have been developed for joining cylinders, wherein the cylinders are rotated past a stationary welding head.

It is almost impossible to restart the process in mid-seam should a breakdown occur, and the quality cannot be easily judged until the mould shoes have cleared the solidified weld metal. In practice, therefore, the process requires a great deal of care and time in setting up, although completion is thereafter rapid.

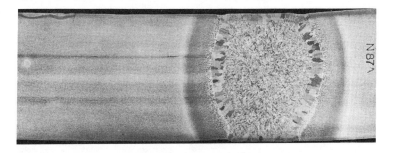

Figure 1.14b   Cross section of electro-slag welded BS 968 steel test plate (3⅜ in thick). Note also hydrogen-induced toe crack caused by careless application of a manual-metal-arc tack weld. (Courtesy: Motherwell Bridge and Engineering Company)

The fluxes used have a different composition from those which have already been described. They are designed to quench arc formation (and not to encourage it) and to have a suitable electrical resistance when molten. Viscosity must be carefully balanced, as the slag should be fluid enough to allow convective heat transfer to the plate edges, but viscous enough to minimise leakage past the sliding copper shoes. For aluminium and titanium joints, the slag bath is insufficient to exclude air, and an extra blanket of inert gas is directed on the slag pool surface.

The metallurgical appearance of electro-slag joints is also unlike other fusion welds (see *Figure 1.14b*). The slow heating and cooling cycle reduces the risk of hydrogen-induced cracking, but allows substantial grain growth. As a consequence, fracture toughness is usually poor. If the completed component is normalised (at about 950 °C for steel) in a furnace, the structure can be refined, but this is often impracticable for large components which cannot support their own weight at these temperatures. Considerable development effort has been put into fine-grain electro-slag techniques, which are largely based on increased welding speeds. If the speed is pushed too high, the risk of centreline cracking is great, as large thermal stresses are generated.

There are many variants of the process, with special attractions and disadvantages. For example, a considerable mechanical simplification is made in the 'consumable nozzle' version. The wire-feed machinery is fixed at the top of the joint, and wire is pushed down a consumable guide tube which shortens by melting as the weld progresses. The guide nozzle can be shaped to suit a curved joint, and simpler leaktight moulds can be used.

The electro-slag process and variants have been applied to pressure-vessel manufacture (mostly longitudinal seams), ship construction (taking advantage of the natural vertical presentation of structural members), and the joining of heavy forgings and castings for turbine parts, valves and the like.

The electron-beam welding process

In this new fusion welding method, heat is generated when a high-velocity focused stream of electrons collides with the workpiece. The interesting feature of this principle is that the energy density at the heated spot can be 5000 times greater than that associated with electric arcs, and very narrow, deeply penetrating welds can be formed at speeds of the order of 30 m/min. Metallurgical damage to adjacent material and thermal distortions are, therefore, greatly reduced. As welding is carried out in a vacuum chamber, the atmosphere is positively excluded.

A typical machine arrangement is shown in *Figure 1.15*. Electrons are emitted from a hot tungsten cathode, and accelerated across the cathode–anode space by a potential difference which is typically in the range 15 to 200 kV. In 'work-accelerated' machines the workpiece forms the anode. This arrangement is only suitable for the welding of components which have a standard size, shape and surface quality, as variations in these factors alter beam focus during welding. More commonly, a 'self-accelerated' machine is used, in which the anode is a perforated disc, and electrons pass through the anode to be focused by electromagnetic coils. Deflection coils are incorporated to allow fine adjustment of the beam impingement spot.

The vacuum not only prevents weld pool contamination, but is vital to the electron beam efficiency. Frequent collisions between the electrons and gas atoms in the path would scatter the beam and its power would be reduced. The vacuum chamber requirement inhibits application of the process for large components, or for repetitive work where chamber evacuation times are a significant proportion of the total production cycle. It is not surprising, therefore, that great efforts are being made to circumvent this problem. Out-of-vacuum techniques have been developed, in which the beam is generated and focused in a high-vacuum region, and is passed through a series of orifices and graded pressure regions to the workpiece which may be at atmospheric pressure and is, in such cases, protected by a separate inert gas shield.

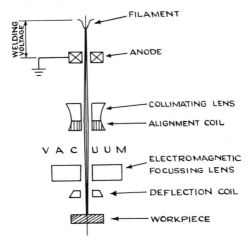

Figure 1.15   Schematic electron-beam welding machine

The beam scatter depends on the atomic number of the shielding gas, and helium which has a lower number than argon, scatters less and allows greater penetration. These techniques obviously require higher accelerating voltages to offset beam scatter.

Characteristics of the electron-beam process

The concentrated nature of the heat source makes the process very suitable for metallurgically difficult metals and dissimilar metal joints. It has been used for fully heat-treated steels, difficult aluminium alloys, titanium, tungsten, beryllium, zirconium, and combinations of these. The vacuum chamber is especially suited to reactive metals. The available power is very controllable, and the same machine can be applied to single-pass welding of stainless steel at 50 mm thickness, or aluminium foil at 0.1 mm thickness. The beam can be projected through narrow openings, to weld in areas of limited access. The very low distortion feature has encouraged its use for the joining of fully machined parts, and the salvaging of machine shop mistakes. Tubes, gear clusters, turbine blades, and heat sensitive electronic component applications have shown the process to best advantage.

Apart from the practical limitations of vacuum chamber operation, the process demands precision and sophistication. Metallurgically clean parent materials and close fits are essential. However, development is still at an early stage, and we can look forward to further interesting applications which exploit the excellent potential of such a high-energy heat source.

Oxy-acetylene welding

In this familiar fusion process, heat is generated by exothermic reaction in an acetylene/oxygen flame. Welding rates are slow, as the flame temperature is low (3500 K) relative to electric arcs, and the heat source is external to the workpiece.

The gas flows are carefully metered, mixed, and delivered to the welding blowpipe tip. Chemical reactions between the flame and the metal depend on the gas volume flow ratio. Equal volumes produce a so-called 'neutral' flame, excess acetylene a carburising flame, and excess oxygen an oxidising flame. Neutral flames are normally used for steels, to minimise alloy losses and maintain the original carbon level. Oxidising flames are used for copper alloys and carburising flames are used for reactive metals such as aluminium where oxygen pick-up would be undesirable. A clean workpiece is essential.

Filler metal may be added separately, as in the GTA process, and chemical fluxes are often used to prevent formation of oxides when aluminium, cast iron, bronze and stainless steel are welded.

The process is versatile, and the equipment is simple, cheap and easily transported. It therefore lends itself to repair and maintenance applications. However, considerable welder skill is required. It is suitable for thin materials or in other work where close control of the weld pool is required, as in tube butt welding. Other applications include defective casting repair and hard surfacing.

Summary

The performance of the principal fusion welding processes is summarised in *Table 1.1* (page 9).

1.2.2 Resistance welding processes

Heat is generated in these processes by electrical resistance heating of the component interfaces, when they are pressed together and a current is passed. The application of pressure through the contact electrodes has several important functions—air is excluded from the weld zone, the electrode-to-workpiece contact resistance is reduced and, in some cases, molten metal and oxide is expelled from the interface to allow solid-state bonding.

Spot welding

The principles of spot welding are illustrated in *Figure 1.16*. The sheets to be welded are clamped between two opposed electrodes, and a high current (up to 100 000 A) is passed for a very short time. A molten pool is formed at the sheet interface, the size of the final weld being closely related to the electrode size.

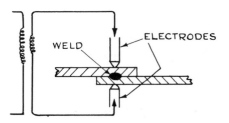

Figure 1.16   The spot-welding process

Many variations in the current/time/pressure sequence can be obtained using modern spot-welding machinery. In particular, pulses of current before and after the welding phase can be used to preheat and temper the materials. If excessive current is used, weld metal may be ejected from the molten zone. This is called 'splashing'. Excessively long welding times can also allow splashing, or arcing and melting at the electrode contact point. Unnecessarily large weld and heat-affected zones will also result.

The electrode diameter used is usually specified according to the sheet thickness by the simple formula $D = \sqrt{t}$, where D is the electrode tip diameter, and t the sheet thickness (for a two-sheet weld). If metals of dissimilar electrical resistance are welded, the molten zone grows preferentially towards the high-resistance side. This can be avoided by specifying a smaller diameter or a higher-resistance electrode on the side having the higher electrical conductivity.

The process is best applied to sheets less than 6 mm thick, and has been used for aluminium, molybdenum, titanium, carbon and low-alloy steels, and some dissimilar metal combinations. High-conductivity materials such as copper and silver are difficult to weld. Impressive multi-head welding systems with automatic loading, transfer and weld sequencing are used in car-body production.

Resistance seam welding

The principles of resistance seam welding are very similar to the previous case. The sheets are nipped between two copper-alloy roller electrodes, which travel along the joint to produce a series of spot welds. This arrangement is illustrated in *Figure 1.17*.

The current pulses and rotational speed of the rollers can be

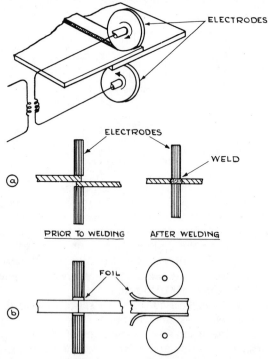

Figure 1.17   The seam welding process and its modifications. (a) Mash welding process; (b) foil–butt seam welding

arranged to produce a series of overlapping spot welds, or separate spot welds (called 'roller-spot' welding). Further variants called 'mash welding', and 'foil-butt welding' are illustrated in *Figure 1.17*.

Projection welding

In this process, small prepared projections on the workpiece surface are melted and collapse when current is supplied through flat copper-alloy electrodes (see *Figure 1.18*). The projections are formed (usually on the thicker or higher-conductivity side of the joint) by pressing.

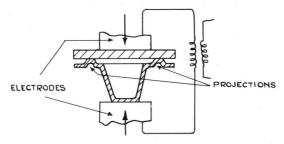

ELECTRODES PROJECTIONS

Figure 1.18   The projection welding process

forging or machining. The projections are shaped and positioned to concentrate the current and a large number of welds can be made simultaneously. Lower currents and pressures are used than for spot welding to avoid collapse of the projections before the opposite surface has melted.

The process is best suited to mass production, and is popular for the welding of nuts and studs to sheet metal sections.

Flash welding

This process is widely used for workshop butt welding of tube lengths, and is illustrated in *Figure 1.19*. The workpieces are clamped in a special jig, and brought lightly together. A high current is transmitted through the clamps and the small areas of contact between the workpieces are quickly superheated until molten metal is ejected. (This is called 'flashing'.) Flashing continues while the workpieces are brought slowly closer, until the whole area of contact is covered, and a molten zone is formed. At this point the current is switched off, and the workpieces are forced together, squeezing out the metal formed during the flashing phase from the joint interface. The forging or upsetting phase disrupts the oxide film and removes contaminated material from the weld zone to allow clean bonding. The 'flash' or contaminated metal is removed after welding, since it is metallurgically unsound and forms a stress-concentrating feature (see p. 202).

The workpiece approach speed, the current, and the flashing time must be carefully controlled if the region which is later upset is to be efficiently heated. Excessive current can cause lack of fusion defects, as craters are formed at the interface during flashing, and are not filled at the forging stage. In the case of tube butt welding where

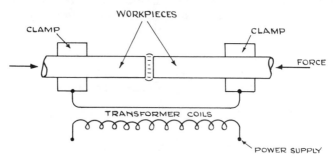

Figure 1.19   The flash welding process

the internal flash is difficult to remove, an internal gas pressure is often applied to minimise bore obstruction. The process is widely used to attach small fittings to larger forgings.

1.2.3   Solid-state bonding processes

It will become even more apparent in Chapter 2, that the high temperatures, melting and rapid cooling which are fundamental to fusion processes can lead to undesirable metallurgical structures which have a poor service performance. Although an apparently wide variety of materials can be fusion welded, the detailed compositions are, in most cases, specially selected to be 'weldable', and a considerable range of common engineering materials fall outside this classification. The problem of dissimilar metal joints is especially acute, as brittle intermetallic compounds are often formed after melting and mixture.

For these reasons the solid-state processes (in which no melting occurs) are especially attractive. The broad principle used is that any oxide films are first broken up and removed, and then the surfaces are pressed together until intimate contact and bonding occurs. Heating is still often used, however, as high temperatures assist the removal of oxide films, allow easy plastic flow, and increase the atomic migration rate; but temperatures are held below the melting point, which is a significant value for metallurgical transformations. To sum up, therefore, the principal advantages of solid-state processes are:

The cast structure associated with fusion welds is avoided

No filler metal additions are required, and the physical and mechanical properties of the joint are similar to those of the parent metal

The size of the weld zone and the thermal distortion are less than in most fusion welding processes

A greater range of materials can be joined without deterioration of physical and mechanical properties. Dissimilar materials can also be bonded (steel/nonferrous, plastics, etc.)

Of course, some of these advantages tend to slip away as a greater amount of heat is used and, in friction welding for example, the combination of high temperature and extreme plastic deformation may be more than the material can stand. Also, the components to be welded must be of a form and size which is convenient for the application of forging forces. This may place practical restrictions on the application scope of solid-state processes relative to fusion welding.

Diffusion bonding

At temperatures between one-half and two-thirds of the melting point, sufficient atomic diffusion occurs for bonds to be made between flat clean surfaces which are pressed together. Air is excluded from the joint interface by an inert gas or a vacuum chamber. The vacuum, if used, also assists the breakdown of oxide films, and allows them to diffuse away from the interface.

In roll bonding, the plates to be welded are stacked in sandwich form, and rolled together. Air is excluded by welding around the edges of the interface to form a seal, or an inert gas or vacuum is used. This method is used commercially in the production of clad plates, where a thin layer of expensive corrosion-resistant metal is bonded to a cheaper thick base material.

The eutectic-diffusion bonding process uses an interesting principle, which is that foreign atoms will diffuse more readily in a material than parent atoms. An intermediate layer of dissimilar metal is therefore placed between the workpieces and diffusion takes place at high temperatures in a vacuum. The layer may be produced by electro-deposition, or may be provided by a metal foil.

Aluminium, titanium, and zirconium and many dissimilar metal combinations can be diffusion bonded. Even ceramic-to-metal joints have been made.

There are, however, practical difficulties in applying the process; namely the need for a vacuum, the need for precise temperature control, and the difficulty in applying pressure uniformly over the whole joint area. Recent published work has shown, however, that steel can be bonded at 1200 °C using pressures as low as $35 \, kN/m^2$ ($5 \, lbf/in^2$). In this application, the periphery of the joint is sealed by welding, and the contacting surfaces are cleaned by solution of the oxide in the steel (so-called 'auto-vac' cleaning). Similar techniques should be applicable to any metal which dissolves its own oxide, and future developments of this technique should be interesting.

Explosive welding

In this process, an explosive charge is used to bring the components together at high velocity. The cladding of plates is illustrated in *Figure 1.20*. The 'flyer' plate impacts the stationary plate at a small angle of incidence. At the interface, the shear stresses are so high

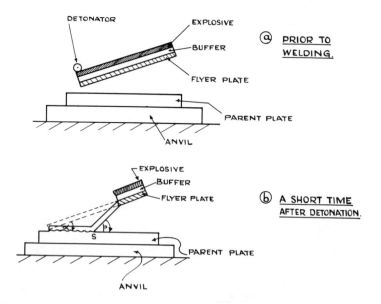

Figure 1.20   The explosive welding process. (Bahrani, A. S., and Crossland, B., Metals and Materials, Feb. 1968)

that the material behaves like a low-viscosity fluid, and the oxide film is swept from the joint to allow clean metal bonding. A characteristically rippled interface is normal. The process has been used for the cladding of titanium, aluminium, zirconium, copper alloys and nickel alloys to a steel base plate.

Ultrasonic welding

The weld is produced in this process by mechanical vibration at ultrasonic frequencies (up to 100 kHz). The sheets to be joined are placed together on an anvil, and an ultrasonic vibrator is applied to the opposite face (see *Figure 1.21*). Mechanical deformation at the

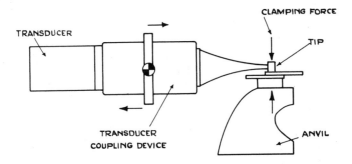

Figure 1.21   The ultrasonic welding process. (After Vernon, C. W. J., *Welding and Metal Fabrication,* Nov. 1965)

sheet interface disrupts the oxide film, and metallic bonding follows.

Although there is no evidence of melting, metallurgical recrystallisation precipitation and phase transformation can occur. For reactive metals such as aluminium and titanium, the temperatures reached are sufficient for oxides to form in the weld region unless inert gas shielding is used.

The welding time is extremely short, and little bulk deformation of the sheets results. The process has been successfully applied to sheets from 0.004 mm to 2.5 mm, in steels, nickel alloys, copper alloys, aluminium, gold, silver and in dissimilar metal combinations.

Friction welding

In this process, the parts to be welded are rubbed together while a pressure is applied to the interface. Frictional forces disrupt the oxide films and generate high temperatures at the interface. Near the end of the welding cycle, the rubbing action is arrested and pressure is increased to upset the workpieces and expel oxidised material. A typical application suitable for rotatable components is illustrated in *Figure 1.22.* Torque measurements on instrumented welding machines reveal a number of stages in the frictional heating process, as the two surfaces rub each other until matching is possible. Eventually the torque required to maintain speed stabilises. The total heat generated is proportional to the square root of pressure. Although the maximum relative motion occurs at the joint periphery, heat conduction produces a sufficiently uniform temperature across

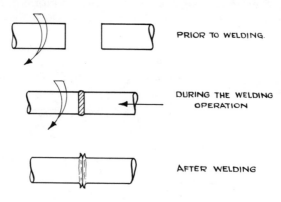

Figure 1.22   The friction welding process

the joint. Careful sequence control of rotational velocity, heating time, initial pressure and final upsetting force and distance are necessary. In one application, known as 'flywheel' or 'inertia' welding, the energy required is stored in a rotating flywheel and control is effected merely through the initial flywheel energy and the interface pressure.

Whether melting occurs or not (and there is still some argument on this point), high-quality welds are made in which the heat-affected zone is small, and the grain structure is refined by the hot mechanical working which is a part of the process. A wide range of materials have been successfully joined using this process, including dissimilar combinations and heat-treatable alloys. Many of the components in the automotive industry are especially suitable (valves, shafts, etc.), and in this respect, the process is something of a competitor with flash welding.

1.3 Understanding weld preparations

The reader who is even slightly familiar with welding procedures will already know that the edges of components to be joined are frequently cut away to form a groove or 'preparation'. This is especially true of thicker sections, and although it may seem illogical to remove sound parent metal and subsequently replace it with expensively deposited weld metal, there are valid reasons for this procedure.

There is no shortage of recommendations for joint preparations and, for convenience, a representative selection is included in the Appendix to this chapter (*Figures 1.42–1.44* and *Table 1.4*). These designs represent the last stages only, of a long process of trial and error; it is therefore difficult to appreciate the individual reasons for particular features, even after careful examination of the details; for example, why is a certain wide angle specified in one case, and a narrow angle in another? . . . and so on. One can understand that a designer who has no recourse to practical welding advice does not find it easy to select suitable designs from the range offered. He will be even more reluctant to modify or devise preparations to suit specific cases. This is usually a pity, for it is impossible to make the best use of available equipment and skills unless the joint preparation is well adapted to the application.

It may be instructive therefore to follow the development of joint details from the simplest case, and to examine some of the arguments for particular features. To save space, the scope of the discussion is confined to electric arc fusion welding.

1.3.1 Heat distribution and penetration control

The central purpose of a joint preparation is to make the best use of the heat input and penetrating force of the chosen process. We might begin by considering the simplest possible arrangement, namely, the 'square close-butt preparation' illustrated in *Figure 1.23a*. The maximum feasible joint thickness depends on the depth of penetration available from the desired process. In practice, the maximum possible penetration may not be applicable for a number of reasons; the associated arc force can blow the molten metal through the bottom of the joint; excessive parent metal dilution of the fused zone may cause metallurgical problems (see Chapter 2); the weld pool size may have to be restricted for positional welding, or to avoid a coarse-grained structure. As a guide therefore to

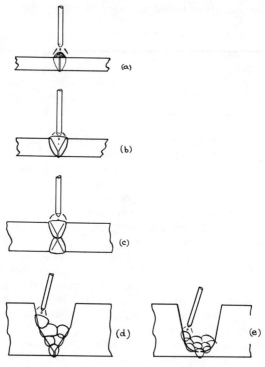

Figure 1.23   Stages in the development of joint preparations

maximum practicable penetration levels for controlled penetration of downhand square close butts, the following limits are stated:

Process	*Max. thickness*
Gas tungsten arc (TIG)	4 mm (2 mm better)
Pulsed or dip transfer gas metal arc	3–4 mm
Constricted plasma arc	up to 12 mm depending on material using 'keyhole' technique
Manual metal arc	2–3 mm

If mechanical support is provided for the weld in the form of a permanent or temporary backing strip, then higher penetration processes can be used with advantage. The shipbuilding industry among others, has a special interest in single-pass welding using such techniques as it can be difficult to turn over large panels in the fabrication shop for two-sided welding.

As a first step towards the welding of thicker sections, an improvement in control can be obtained if the plate edges are bevelled (*Figure 1.23b*). This simple modification allows a reduction in the arc force necessary to penetrate, and concentrates the available heat in the area to be melted. A one or two millimetre root thickness or 'nose' is retained in the preparation, as a sharp edge tends to burn away.

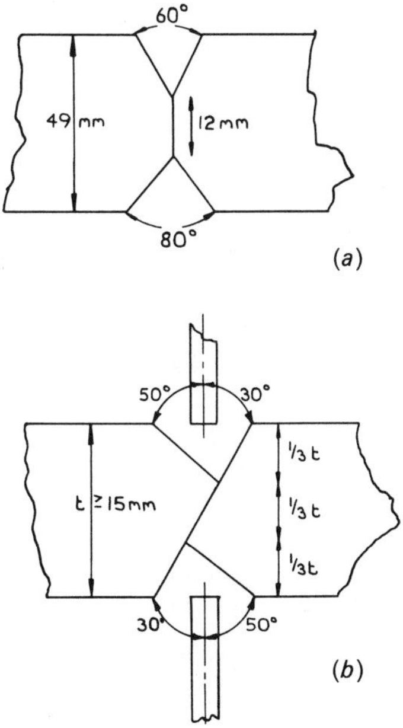

Figure 1.24   Standard (a), and K–Z (b) preparations for submerged-arc welding (after Maeda)

The bevel may be combined with (or less satisfactorily replaced by) a gap, which permits increased penetration and a thicker nose. However, automatic welding applications are less tolerant of gap variations due to poor fitting or distortion during welding, and it may be wiser in such cases to do without a gap. (Maeda has reported the use of an ingenious 'K–Z' preparation to circumvent this problem. This preparation is illustrated in *Figure 1.24* and allows a fitting tolerance without introducing a troublesome variable gap.)

The choice of bevel angle is essentially a compromise between a large angle for control, and a small angle to minimise the volume of deposited metal.

1.3.2 Multi-pass techniques

For greater thicknesses it is clear that several weld runs will be needed. This has some incidental advantage, in that each run partially refines the cast grain structure of the previous one. If the component can be conveniently turned over, the joint may be attacked from both sides. It is important then to make sure that the opposing fused zones interpenetrate soundly with no entrapment of slag or other defects. The square close-butt preparation is suitable, but the addition of simple grooves assists penetration and helps to stabilise the arc position to the centreline of the joint (see *Figure 1.23c*). In careful practice, the groove on the second side may be cut *after* welding the first side to remove any root defects which might be subsequently trapped. This is known as 'back gouging'.

When the joint thickness exceeds that which can be comfortably completed in three or four runs per side, it is worth introducing a further modification, by steepening the sides of the preparation above the root area. This of course reduces the volume of metal required to complete the joint as shown in *Figure 1.23d*. Completely vertical sides are not recommended, however, as the electrode needs to be angled to the wall in order to penetrate soundly. The angle transition from the root should not be abrupt (i.e. not greater than 25° in one step) as there is then a risk of crevice formation at the corner. Trapped slag or incomplete penetration could result.

At this stage of development for thicker joints there is no need to compromise on root angle, and as an alternative to the previous design, the angle can be increased to 180° and radiused smoothly into the 10° walls as in *Figure 1.23e*. This so-called 'J' or 'U' preparation is obligatory for GTA root runs, as surface tension effects in a V shape tend to suck material from the underside of the joint to form a 'sink' or depression there.

It might be mentioned incidentally that a very narrow preparation can cause centreline cracking, if the weld bead is forced into a deep narrow shape by the constriction. Large root gaps can also have a direct bearing on root cracking, particularly in heavily restrained joints or in other examples where hydrogen-induced cracking is possible (see Chapter 2).

1.3.3 Joining thick to thin thicknesses

The essential problem in joining unequal thicknesses is one of heat balance. When sufficient heat is provided to fuse the thicker side of the joint where heat removal is rapid, the thin side overheats and may burn away. A better balance is obtained by locally thinning the thick side of the joint or by arranging the joint so that the thin material is protected by a heat sink. Both of these methods are illustrated in the recommended tube-to-tubeplate details shown in the Appendix to this chapter (*Figure 1.43*).

1.3.4 Access

There is the story, popular among welding foremen, of the unfortunate designer of a metal box, who, unwittingly, required a welder to be incarcerated in each of his products. The story is no doubt apocryphal but welds are still occasionally specified in places where the welder has insufficient room to move and has a poor sight of the joint.

The joint preparation itself must be wide enough to allow manipulation of a covered electrode or a welding gun. Positional welding requires even greater flexibility of movement and it is usual to find that wide angles are specified for such joints. An asymmetrical preparation is often used for horizontal-vertical joints, as it helps the welder to build successive runs against gravity.

Automatic processes may show an advantage in allowing narrower preparations, provided that the welding head itself is not too bulky and the electrode can be satisfactorily positioned.

1.3.5 Distortion

It does not anticipate the next section too much to observe at this stage that the choice of joint preparation has a considerable effect on welding distortion. Aside from in-plane longitudinal and transverse contractions of the joint, the angle between the components tends to reduce towards the side on which the heat source is located (angular distortion). In general terms, distortion of all kinds increases with the volume of metal deposited. It follows that a preparation which minimises the volume, for example, a 'U' or 'double-V' detail instead of a 'single V', offers a double bonus in economy and minimum distortion.

If double-sided welding is employed, the angular distortion can be balanced on alternate welds. The movement is always greater for the earlier weld runs where the components are free, and it may be necessary to use an asymmetrical preparation (smaller volume on the first side to be welded) or to counterbalance the first run with two runs on the second side, and so on for further layers. An asymmetrical preparation also allows the first run of a three-or-more run application to be placed on the neutral axis.

1.3.6 Lamellar (or laminar) tearing

In rolled plate, it is not unusual to find areas of weakness, caused by the clustering of nonmetallic particles in layers parallel with the plate surfaces. These faults are difficult to detect nondestructively. They

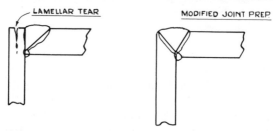

Figure 1.25a

Figure 1.25b   A typical subsurface lamellar tear

cause little reduction in mechanical performance, unless loading is applied in the thickness direction as it may be in corner, T joint, or cruciform configurations. In these cases especially, thermal stresses are generated in the thickness direction during welding, and the weak areas may be torn open (see *Figures 1.25a* and *b*).

To avoid this possibility, the designer may specify cleaner material, at presumably greater cost, or design the joint to replace suspect areas with weld metal which is hopefully sound.

1.3.7 Summary

The weld joint should be located and shaped to be comfortably accessible in terms of the desired process and welding position. The detailed shape is designed to distribute the available heat adequately, and to assist control of penetration. The volume of weld metal should be minimised to reduce cost and distortion.

It is clear that many common welding faults can be directly or indirectly traced to poor specification of the weld preparation. It is therefore wise to consult the welding personnel concerned at an early stage, if subsequent recriminations are to be avoided.

1.4 Prediction and control of distortion

The contractions and distortions which accompany thermal welding and cutting processes can create considerable problems, particularly for fabricators of thin plate structures. The root cause is the nonuniform temperature field associated with these processes. The resulting thermal strain differences cannot be accommodated without exceeding the elastic limit and permanent strains and stresses are left behind after the heat source has passed over the component. The main difficulty lies in predicting the final size and shape of the fabrication, given specific initial dimensions. Post-weld correction of distortion by machining can be a frustrating procedure, because removal of stressed material alters the internal equilibrium of the component causing further distortion.

The qualitative features of typical welding distortions can be understood by considering the effect of contractions of the intensely heated zone in directions parallel and perpendicular to the direction of travel of the heat source. These cause *in-plane contractions* in the first instance (*Figure 1.26*). If the weld line does not coincide with the

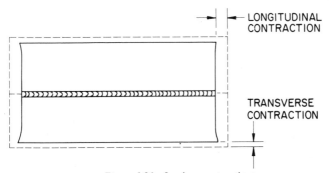

LONGITUDINAL CONTRACTION

TRANSVERSE CONTRACTION

Figure 1.26   In-plane contractions

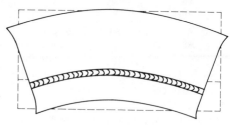

Figure 1.27 In-plane curvature

neutral axis for bending, *in-plane curvature* will result (*Figure 1.27*). Also if the through-thickness temperature pattern is not symmetrical with respect to the neutral axis of the component, transverse and longitudinal *out-of-plane curvatures* will result (*Figure 1.28*). Transverse curvature is often localised to the weld area and is then better described as *angular contraction*. These out-of-plane distortions are of particular significance with respect to buckling tendency (see Section 4.1.9) and are also responsible for the 'hungry horse' appearance of stiffened plate structures such as ships. All forms of contraction and distortion tend to increase with the cross-sectional area of the fused zone (other things being equal) and therefore in most practical cases the heat input per unit length of weld ought to be minimised if the aim is to reduce distortion.

Most of the available information and advice on the subject is concerned with qualitative techniques and good practice in reducing distortion[1,2]. For example, the sequence of fabrication can sometimes be chosen to avoid welding off the neutral axis (*Figure 1.29*) or to make sure that the contracting weld is presented with the maximum

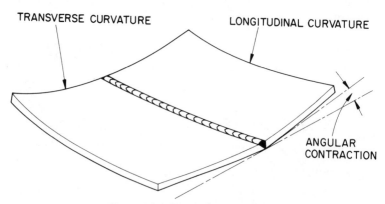

Figure 1.28 Out-of-plane distortions

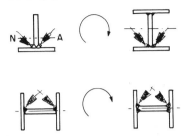

THIS SEQUENCE GIVES LESS OUT-OF-PLANE DISTORTION
Figure 1.29

structural stiffness. Some compensation for angular contraction can be incorporated by presetting the assembled components in a direction opposite to the contraction produced by the weld. However, for those who require a specific numerical prediction of contraction and distortion, or who wish to establish the relative merits of different welding sequences or procedures, or who are tackling different materials, there is little information to assist. In 1955 a Soviet researcher Okerblom[3], established a set of simple analyses which cover a number of common distortion modes and although experimental confirmation of these analyses is meagre, they provide (with some extensions in the present text) a useful framework for the numerical prediction of contraction and distortion.

Okerblom's theory is based on a two-dimensional heat flow treatment, in conjunction with a one-dimensional stress analysis. An elastic/perfectly-plastic stress–strain relation is assumed for the material (see Section 4.1.1) and above a given 'softening' temperature, it is supposed to exhibit zero yield strength. The expansivity coefficient α is also assumed to be constant at all temperatures. The introductory discussion of heat flow theory which follows is also relevant to the later chapter on the metallurgy of welds.

1.4.1 Temperature fields

The temperature distribution associated with a moving heat source has been established, apparently independently, by Rosenthal[4] and Rykalin[5]. The important finding of these solutions is that the temperature is constant at fixed distances from the heat source, although at a specified point in the component, the temperature cycles up and down as the arc passes. Thus for given materials and welding parameters, the heat-flow solution can be expressed in terms of a stationary pattern of isotherms where the plate is moved through the temperature field at a constant velocity corresponding to the weld travel speed. For high-energy, rapidly moving heat sources, the solutions can be simplified to show that the isothermal contours depend principally on the following parameters:

q heat input rate, equal to power consumed in the arc, less losses (W or J/s)
v welding speed (m/s)
t plate thickness (m)
K thermal conductivity of the material
$c\rho$ volumetric thermal capacity (J/m³ degC)

The latter two properties are sometimes combined to give a third:

$$\lambda = K/c\rho$$

which is called the *diffusivity* and measures the spreading rate of a temperature field (m²/s). *Figure 1.30* gives an example of such a temperature field.

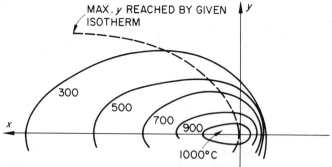

Figure 1.30   *Two-dimensional quasi-static temperature field (after Rosenthal[4])*

Typical thermal properties of common metals are given in *Table 1.3*.

Certain features of the theoretical temperature patterns are worth noting: the heated zone normally takes the form of a narrow strip, so that the thermo-mechanical effect on the plate is similar to that which would be produced by simultaneous heating and subsequent cooling of a narrow strip (as defined by the 300 °C isotherm, for example). The maximum width of each isotherm is closely related to the heat input per unit length times thickness (q/vt); in fact for typical high-energy heat sources the maximum width w of an isotherm at temperature T is given approximately by

$$w = \frac{1}{2}\frac{q}{vt}\frac{1}{c\rho}\frac{1}{T} - \frac{4}{5}\frac{\lambda}{v} \tag{1.1}$$

Table 1.3 THERMAL PROPERTIES OF COMMON METALS
(AT ABOUT $\frac{1}{3}$ OF MELTING POINT WHERE POSSIBLE)

	Coefficient of expansion α (degC^{-1}) $\times 10^{-6}$	Volumetric thermal capacity (J/m^3 °C) $c\rho$ $\times 10^{+6}$	Approx. distortion factor (m^3/J)(see App. Note 1) $\alpha/c\rho$ $\times 10^{-12}$	Diffusivity $\lambda = K/c\rho$ (m^2/s) $\times 10^{-6}$	Melting point (°C)
Aluminium	23–27	2.7	8.5–10	85–100	660
Carbon steel	14	4.5	3.1	9.1	1400
9% Ni steel	13	3.2	4.1	11.0	1400
Austenitic steel	17	4.7	3.6	5.3	1500
Inconel 600 (75% Ni, 15% Cr)	14	3.9	3.6	4.7	1400
Titanium alloy	17	3.0	5.7	9.0	1650
Copper	17	4.0	4.3	96.0	1050
Monel 400	16	4.4	3.7	8.0	1300

See also: Welding Production, **19**(1). (Jan. 1972); *American Institute of Physics Handbook*. McGraw-Hill.

The second term will be insignificant for rapidly-moving high heat-input rate processes applied to material of low diffusivity. (Okerblom disregards it entirely for welding of steel.) Equation (1.1) also characterises the envelope of the temperature profiles transverse to the weld line at various distances from the source (see *Figure 1.31*). Note however that there are large differences in the thermal patterns associated with materials of different diffusivity subject to the same welding parameters (see *Figure 1.32*).

1.4.2 Longitudinal contraction due to a single weld pass

The thermal stress and deformation patterns are much more difficult to calculate than the temperatures and Okerblom simplifies matters considerably by considering only the elastic/plastic strain cycle which would be experienced by a transverse strip of material passing through the temperature field (hence 'one-dimensional' stress analysis). He also assumes transverse cross sections remain straight (plane sections remain plane as discussed later in Section 4.1.2) and that there is no external load or moment. The details of Okerblom's method are not particularly easy to follow and a simpler exposition

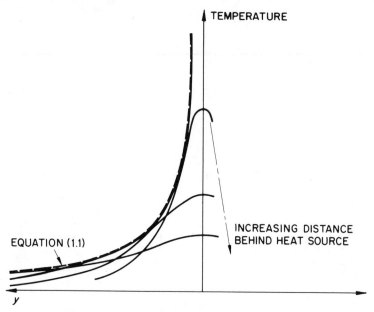

Figure 1.31   Temperature profiles transverse to weld line

based on similar assumptions and leading to the same conclusion is given in Wickramasinghe and Gray[6]. This shows that the unbalanced longitudinal contraction force developed as a consequence of thermal cycling is given by

$$F_l = \tfrac{1}{2}E\,\frac{q}{v}\,\frac{\alpha}{c\rho}\ln 2 \tag{1.2}$$

where E is Young's modulus and the diffusivity correction term is neglected, and the resulting longitudinal contraction strain is given by

$$\varepsilon_i = 0.35\,\frac{q}{v}\,\frac{\alpha}{c\rho}\,\frac{1}{A} \tag{1.3}$$

where A is the total cross-sectional area resisting contraction. If the diffusivity correction factor is included, a term given by

$$0.8\,\frac{\lambda}{v}\,\frac{t}{A}\,\varepsilon_{\text{YIELD}}$$

is subtracted from (1.3). A fortunate consequence of the analysis is that provided that the free thermal strain associated with the 'softening' temperature is greater than twice the yield strain, this temperature

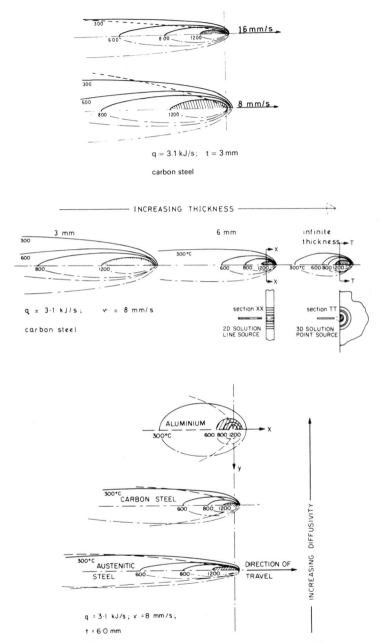

Figure 1.32 Effect of thermal parameters and welding variables

and the yield strength drop out of the calculation, although obviously the yield strength affects the final residual stress state.

If the weld line is offset from the neutral axis of the structural section the contraction force given by equation (1.2) will bend the section, whereupon the in-plane curvature will be given by

$$1/R = \varepsilon_l z A/I \tag{1.4}$$

R being the radius of curvature, z being the distance of the weld line from the neutral axis and I the second momemt of area of the whole cross section (after welding).

Experimental confirmation of these relationships is somewhat lacking; Okerblom quotes one set of test results on steel without much description of the experimental details, and the present writer has produced some experimental evidence from tests carried out on aluminium plate welding[6]. These at least confirm the order of magnitude of the equations, and also demonstrate the importance of the two leading terms q/v (the heat input per unit length of weld) and the thermal distortion factor $\alpha/c\rho$ which is included in *Table 1.3* to give a quick guide to the distortion propensity of different metals. Notice however that at low travel speeds such as occur in manual TIG welding, the diffusivity factor becomes important and probably accounts for the difference in distortion tendency between austenitic and carbon steel.

The following example will help to clarify the method of calculation for in-plane contractions and curvatures due to a single weld pass.

Example 1.1

Fifteen-metre lengths of steel tube and fin are welded together by the CO_2 dip-transfer process (see *Figure 1.33*). The welding parameters are 24 V, 160 A and 8 mm/s. Assuming an efficiency of heat transfer of 80%, find the resulting distortions.

Answer:

$$\frac{q}{vA} \times \frac{\alpha}{c\rho} = \frac{24 \times 160 \times 0.8}{8 \times 1100} \times \frac{3.1}{10^{12}} \times 10^9 = \frac{1086}{10^6}$$

Hence from Equation (1.1), $\varepsilon_l = 362 \times 10^{-6}$

therefore overall contraction $= 15 \times 362 \times 10^{-6} = 5.4$ mm

and curvature $\dfrac{1}{R} = \dfrac{\varepsilon_l A z}{I} = \dfrac{362}{10^6} \times \dfrac{1100 \times 20}{330 \times 10^3}$

$$= 24.2 \times 10^{-6} \text{ mm}^{-1}$$

EXAMPLE I.I

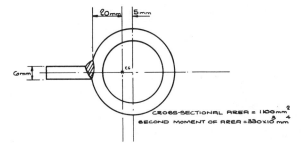

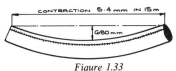

Figure 1.33

From the geometry of shallow curvature, the central transverse deflection relative to the ends of a line with curvature $1/R$, and length l, is given by

$$\delta = \frac{1}{R} \frac{l^2}{8}$$

hence

$$\delta = \frac{24.2}{10^6} \times \frac{15^2 \times 10^6}{8} = 680 \text{ mm}$$

1.4.3. Transverse contractions

To calculate the contractions transverse to the weld it is necessary to consider the relative deformation of the transverse strip in a direction normal to the weld line. However, as the free thermal strains of each element are equal in all directions, the net contraction transverse to the weld ought to be the same as calculated for the longitudinal direction, with the difference that in this case the contraction is concentrated in the weld region and not averaged over the width. Therefore in the case of a sufficiently long weld, the transverse *absolute* contraction will be given by

$$\Delta_z = 0.35 \frac{q}{vt} \frac{\alpha}{c\rho} - 0.8 \frac{\lambda}{v} \varepsilon_{\text{YIELD}} \qquad (1.5)$$

In the case of a weld which completely crosses a section of width B the transverse absolute contraction from equation (1.5) could be written

(neglecting the correction term)

$$\Delta_z = 0.35 \frac{q}{v} \frac{\alpha}{c\rho} \frac{B}{tB}$$

which is the same absolute contraction as would be indicated by the longitudinal contraction formula (1.3) for a weld length B. This discovery led Okerblom to suggest that a short transverse weld could be replaced in an analysis of deformations by an equivalent longitudinal weld of the same length (*Figure 1.34*). This is illustrated in the following example.

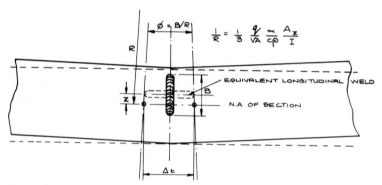

Figure 1.34   Construction and in-plate curvature due to a short single-pass transverse weld

Example 1.2

Estimate the distortion when stiffeners are welded to an I-beam as shown in *Figure 1.35*.

Answer:
Assuming that the four fillet welds can be considered to be applied simultaneously, the heat input q/v may be estimated from the fillet weld size by *Figure 2.16*, as: $q/v = 4 \times 0.8$ kJ/mm. Not all of this heat input goes towards heating and therefore distorting the beam, however. We might assume that the heat flow is proportioned according to the thicknesses of material presented to the heat source, the heat proportion flowing to the beam being given by $10/(10 + 8) = 0.56$. Hence the longitudinal contraction per stiffener is given by

$$\Delta_l = -0.35 \frac{q}{vA} \frac{\alpha}{c\rho} \times 150 = -0.35 \times \frac{4 \times 0.8 \times 3.1}{6 \times 10^3} \times 150 \times 0.56$$

$$= 0.05 \text{ mm}$$

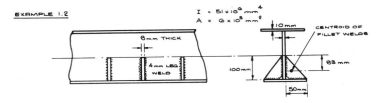

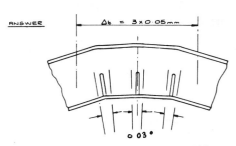

Figure 1.35

and the curvature

$$\frac{1}{R} = 0.35 \frac{q}{vA} \frac{\alpha}{c\rho} \frac{Az}{I}$$

$$= 0.35 \times \frac{4 \times 0.8 \times 10^3 \times 0.56}{6 \times 10^3} \times \frac{3.1}{10^3} \times \frac{6 \times 10^3 \times 83}{51 \times 10^6}$$

Rotation angle

$$\phi = \frac{1}{R} l = \frac{3}{10^6} \times 150 \times \frac{180}{\pi} = 0.03°$$

It should be emphasised that there is no specific confirmation of equation (1.5) and one would expect transverse contractions to be influenced very much in practice by clamping or the restraint of other parts of the structure, particularly if they are stiff and close to the weld line. It may be more realistic to treat the equations as an 'order of magnitude' indicator.

1.4.4 Angular distortion

The development of angular distortion is a more localised process than longitudinal or transverse contraction and it is simply not practicable to follow through the previous kind of thermal/strain analysis in the three-dimensional configurations which are relevant to angular movements. Nevertheless, a simple but useful description of angular contraction can be generated.

It is reasonable to argue that angular contraction is caused in the first instance by through-thickness differences in the isotherm pattern. Realising further that contraction in the other modes is proportional to the width of isotherms and assuming that the isotherm pattern is related at each point through the thickness to the width of fusion zone at that level, it could be concluded that the angular contraction will depend solely on the shape and relative cooling contraction at different levels in the fusion zone. Thus for an unrestrained V-butt weld carried out in one pass (*Figure 1.36a*) the

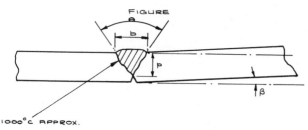

Figure 1.36a   Angular distortion

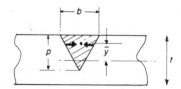

Figure 1.36b   Model for angular contraction

contraction angle β will be given by the difference netween the absolute contractions at the top and the bottom of the V, hence

$$\beta = \tan^{-1}\left(\frac{\alpha T b}{p}\right)$$

or if the fusion zone shape is defined in terms of a weld prep. angle θ:

$$\beta = \tan^{-1}(\alpha T \times 2\tan(\theta/2)) \qquad (1.6)$$

Hence angular distortion will be less when a deeply-penetrating narrow-angled weld zone is produced, either through use of an appropriate process or by choice of weld preparation. Conversely, in the case of surface fillet welds ($\theta = 90°$), the angular distortion will be much greater (this conclusion certainly coincides with experience).

There is therefore a clear moral here for the designer. Of course, equation (1.6) is of limited applicability without a value for the temperature T before cooling and Okerblom recommends 1000 °C for the reason that it agrees with his experiments.

Equation (1.6) needs to be altered, however, if there is some restraint to angular movement; for example, if the plate is clamped to a stiff jig, or a partially-penetrating weld is added to a joint which already has some transverse bending strength. An approximate model of some of the important effects in 'restrained' angular distortion can be developed as follows: first assume that the resultant of the transverse contraction forces coincides with the centroid of the fusion zone, and that yield point magnitude stresses are developed. Hence, the moment causing angular contraction in the case of a triangular section weld (see *Figure 1.36b*) is

$$M = \sigma_Y p \bar{y}$$

If the section behaves like a linear elastic beam (see Section 4.1.2) the curvature in the distorting region will be given by

$$\frac{1}{R} = \frac{\varepsilon_Y}{I} p \bar{y}$$

where I is the second moment of area of the section. The transverse width over which this curvature will take place is difficult to determine but one might assume that it is proportional to the surface width of the weld b. The isothermal contour which gives a 'free' thermal contraction strain corresponding to the maximum elastic range of the material ($2\varepsilon_Y$) seems a dimensionally plausible choice. The ratio between the maximum width of this contour and the fusion zone width b is therefore given by $\alpha T_m / 2\varepsilon_Y$ where T_m is the melting temperature; hence the distortion angle becomes

$$\beta = p b \bar{y} \alpha T_m / 2I \tag{1.7}$$

(This result is almost identical dimensionally and numerically to Okerblom's equation for a parabolic shape of weld, although the derivation was considerably different. Bearing in mind the crude assumptions it would be wise to treat the equation as indicating dimensional trends rather than the absolute value of distortion. Okerblom states, without supporting evidence, that the angular contraction associated with infinitely long welds is three times greater than given above.)

Notice that certain thermal properties which featured in previous distortion equations are not immediately apparent in equation (1.7). This arises because the heated area has in this case been defined directly in terms of the depth and width of the heated zone. Okerblom gives an empirical relationship between heat input and area of fusion in steel welds which shows that $pb = 1.5 \times 10^{-6}(q/v)/(c\rho)$ (J, metric units) and if this is inserted in equation (1.7), it is shown that angular distortion also is proportional to $(q/v)(\alpha/c\rho)$. Therefore the important measures in reducing angular distortion are, reduce heat input per unit length or weld metal volume (as before), keep the fusion zone as narrow-angled as possible and locate its centroid as near as possible to the section centroid. These points are illustrated in *Figure 1.37*.

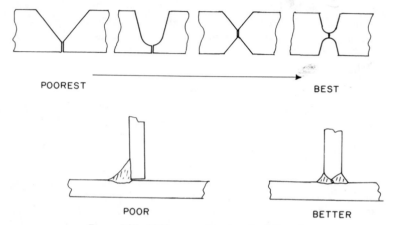

POOREST   BEST

POOR   BETTER

Figure 1.37   Weld preparations to reduce distortion

Example 1.3

An asymmetrical double-V butt weld $(\frac{1}{3} : \frac{2}{3}; 60°$ included angle) is made in a steel plate in two passes (small side first). Estimate the distortion.

Answer:
If the first pass is unrestrained, the contraction angle is given by $\alpha T_m \times 2 \tan 30°$ (equation (1.6)). The contraction in the second pass in an opposite sense to the first pass is given by

$$\beta = \tfrac{2}{3}t(2 \times \tfrac{2}{3}t \tan 30°)(\tfrac{1}{2}t - \tfrac{1}{3} \times \tfrac{2}{3}t)\alpha T_m/(2 \times \tfrac{1}{12}t^3)$$
$$= 1.48\alpha T_m \tan 30°$$

Hence the residual angle is in the first sense, and is given by

$$0.52\alpha T_m \tan 30° = 0.52 \times 14 \times 10^{-6} \times 1000 \tan 30° \times 180/\pi$$
$$= 0.24°$$

This example highlights a general principle that angular distortion tends to be greater on the first pass. If this can be offset by presetting the plates or by clamping, then a large proportion of the final distortion may be avoided. Also, sometimes asymmetrical preparations of this kind are used in three- or four-run procedures whereby the first run is placed on the 'big' side close to the section centroid, and subsequent runs are alternated to 'balance' the welding. Other useful information on angular distortion is given by Masubuchi[7].

1.4.5 Multiple welds

Many components incorporate several welds, applied simultaneously or consecutively and it is important to be able to determine whether the predictions given earlier for single-pass welds still apply.

If the multiple welds are applied simultaneously and at the same travel speed, the effect is simply that the specific heat input is increased and therefore the total heat input can be assumed to supply an equivalent weld which is placed at the centroid of the heat sources.

If, on the other hand, the welds are applied consecutively, it must be appreciated that the second weld is being added to a member which is already stressed and this may alter subsequent deformations. This principle also applies when welding on a member which contains residual stress systems from other operations such as plastic forming.

Consider first the case where a weld is made sufficiently far from preceding welds to be outwith the zones of yield point tensile residual stress associated with these (say greater than $(q/vt)(\alpha/c\rho)(1/\varepsilon_Y)$. If the *initial* stress at the intended weld site is tensile, the plastic contraction is reduced from the stress-free case (see *Figure 1.38*). Conversely if the initial stress is compressive, the final distortion will be greater. Okerblom gives

$$\varepsilon_l^* = \varepsilon_l\left(\frac{\varepsilon_Y - \varepsilon_i}{\varepsilon_Y + 0.7\varepsilon_i}\right) \tag{1.8}$$

where ε_l^* is the strain resulting from applying a weld which would normally give a contraction of ε_l to a site where the initial strain is ε_i. Thus, identical heat-input weld runs do not necessarily give additive or cancelling distortions. This is of considerable significance with respect to the suppression of in- or out-of-plane curvature, as the following case will show.

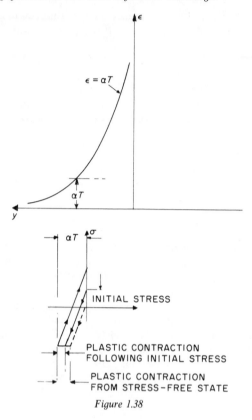

Figure 1.38

Example 1.4

A symmetrical fabricated I-beam is tacked together and high heat-input fillet welds are applied consecutively to the web/flange junctions. Predict the final shape of the beam.

Answer:
Assuming that the beam develops full bending stiffness from the beginning, application of a weld at one junction will lead to a contraction $\varepsilon_l = 0.35(q/v)\ (q/v)(\alpha/c\rho)/A$ at this weld and a residual tensile strain $\varepsilon_i = \varepsilon_l(Az^2/I - 1)$ at the site of the other weld. The initial curvature will be $\varepsilon_l zA/I$ (see *Figure 1.39*).

On application of the second weld, the contraction ε_l at this weld will be given by equation (1.8). Assuming typical parameters for welding steel such that

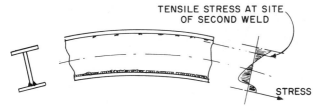

Figure 1.39 *Curvature after first weld on I-beam*

$$\frac{\varepsilon_Y}{0.35(q/v)(\alpha/c\rho)} = 460 \text{ m}^{-2}$$

and section dimensions $A = 6.6 \times 10^{-3}$ m^2, $I/z^2 = 5 \times 10^{-3}$ m^2

$$\varepsilon_l = 0.33\varepsilon_Y$$

hence

$$\varepsilon_i = 0.33\varepsilon_Y\left(\frac{6.6}{5} - 1\right) = 0.106\varepsilon_Y$$

and for the second weld

$$\varepsilon_l^* = 0.832\varepsilon_l$$

Also the opposite curvature induced by the second weld will be less than that induced by the first and the beam will not become straight despite the symmetrical disposition of the welds.

If the distance between consecutive weld centrelines is less than $\frac{1}{2}(q/vt)(\alpha/c\rho)(\frac{1}{2}\varepsilon_Y)$, then the yield point tensile residual stress systems overlap and in the limit, if they overlap identically (equivalent to a re-fusion of the same run) no further contraction will occur, to a first approximation. This situation is difficult to treat in general, but the multipass weld, which is the most obvious practical example of such a case, can be treated reasonably enough by assuming that yield point residual stress is generated within the fused area and to a greater or lesser degree on either side depending on heat-input rate. Thus, for example, in *Figure 1.40* the longitudinal contraction will be given by

$$\varepsilon_l = \varepsilon_Y \frac{\text{Area hatched}}{A} + 0.35 \sum_A^B \frac{q}{vA} \frac{\alpha}{c\rho}$$

where the heat input of the edge runs A to B is aggregated. (Note that for this kind of example where the cross section of an individual weld run is small compared with the total weld cross section, it becomes increasingly difficult to uphold the fiction of a two-dimensional heat-flow theory.) It will be shown in Chapter 2 that an approximate estimate for the size of a semicircular thermal distribution can be

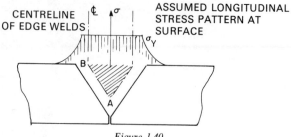

Figure 1.40

obtained which leads to a revised longitudinal contraction, neglecting the diffusivity term, of

$$\varepsilon_l = 0.27 \frac{q}{v} \frac{\alpha}{c\rho} \frac{1}{A}$$

Example 1.5

A single V-butt weld is carried out in six runs with the average heat input for each run being approximately 3 kJ/mm (see *Figure 1.41*).

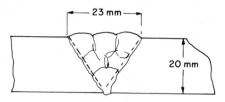

Figure 1.41

Find the longitudinal distortion in a plate 2 m wide if the plate material is steel with a yield strength of 350 mN/m² and a modulus of 200 GN/m².

Answer:
The area of fusion within the approximate centreline of the edge welds is given by $A_f = \frac{1}{2} \times 15 \times 16 = 120$ mm². The tensile residual stress field outwith this area is accounted for by the heat input of three weld runs. Hence the longitudinal force developed is

$$F = \sigma_Y A_f + E \times 0.27n\left(\frac{q}{v}\right)\left(\frac{\alpha}{c\rho}\right)$$

$$= \sigma_Y \times 120 \times 10^{-6} + E \times 0.27 \times 3 \times 10^6 \times 3.1 \times 10^{-12}$$

$$= 0.042 + 1.507 \text{ MN}$$

Therefore longitudinal contraction is given by

$$\varepsilon_l = \frac{F}{AE} = \frac{1.549}{(20 \times 2/10^3) \times 200 \times 10^3}$$

and the longitudinal curvature is given by

$$\frac{1}{R} = \varepsilon_l \frac{zA}{I}$$

$$= \frac{1.93 \times 10^{-6} \times (10 - 20/3) \times 10^{-3} \times 40 \times 10^{-3}}{(2 \times 20^3 \times 10^{-9}/12)}$$

$$= 0.0193 \text{ m}^{-1}$$

(Note that this degree of curvature is unlikely to be developed in practice as the plate is too wide for simple bending theory to apply—see Section 4.1.4.) This also shows that the distortion is not as great as would be expected for the heat input associated with six runs.

It can be seen therefore that for a given cross-sectional area of weld preparation, distortion can only be reduced marginally by restricting the specific heat input so that the sphere of influence of the tensile stress field extends as little outwith the weld preparation area as possible. This can be achieved to a limited extent using a given process by specifying a greater number of low heat-input weld runs, but as we shall see later, such a recommendation could lead to cracking problems and will in any case reduce productivity.

1.4.6 Buckling

Although the compressive residual stress introduced by welding is usually small, it may be sufficient to induce buckling distortion in thin plate structures; either during welding or subsequently, if compressive loads are applied. The likelihood of buckling in a given case may be judged by comparing the residual longitudinal compressive stress ($\varepsilon_l \times E$) with the uniform compressive stress required to buckle the given configuration of plate. Standard solutions covering various shapes of plate and edge-fixing conditions are widely available (e.g. Bulson[8]).

For example, a square plate of length a, simply supported on four sides will buckle under a compressive load given by

$$\sigma = \tfrac{1}{3}\pi^2 \, \frac{E}{1 - v^2} \, \frac{t^2}{a^2}$$

Hence, buckling will occur during welding when

$$0.35 \, \frac{q}{vat} \, \frac{\alpha}{c\rho} \, E = \frac{\pi^2 E}{3(1 - v^2)} \, \frac{t^2}{a^2}$$

i.e. when

$$\frac{q}{v} \, \frac{\alpha}{c\rho} = 0.95 \, \frac{\pi^2}{1 - v^2} \, \frac{t^3}{a}$$

1.4.7 Control of distortion

The numerical treatments which have been given in this section are approximate, but they certainly reflect the important aspects of welding distortion in terms of the parameters drawn into the various equations. However, it would be unfortunate if preoccupation with numbers and formulae obscured some simple understandings of the problem which could help the designer to facilitate fabrication. This final section therefore draws together the points previously made, with the acceptance in the background that distortion cannot be avoided entirely, where fusion welding is concerned, and it is incumbent on the designer to foresee the difficulties which might arise, even if exact calculations are not applicable.

The first and most important point is that all forms of distortion increase with the specific heat input, that is with the heat per unit length required to complete the joint. Therefore processes which minimise the heat input, and generally also the cross section of fused metal, will give less distortion. Thus the electron beam process might represent one extreme of low distortion, with oxy-acetylene welding or submerged-arc welding using wide-angle preparations at the other. For a given process, the designer can also help greatly by reducing the volume of weld metal to be deposited, for example, by specifying butt welds in a prepared groove rather than surface fillet welds, and by using U-grooves rather than V-grooves (see *Figure 1.37*).

The material to be welded is also important especially through the approximate distortion factor $\alpha/c\rho$ and the designer should anticipate greater difficulties with thin aluminium structures for example. The effect of diffusivity is important at low travel speeds and distortion of materials with a high diffusivity (aluminium and copper) may be relatively less if a low speed is being used.

Having made the best attempt to reduce the contraction force developed in a joint, one needs to consider the ability of the fabrication to resist this force. In the case of longitudinal contraction, the axial stiffness is important (proportional to the cross-sectional area) but for the out-of-plane distortions, which by and large are more troublesome in fabrication and assembly, one must try to maximise the bending stiffness and to bring the centroid of the weld as close to the neutral axes of the fabrication as possible. Angular distortion is also reduced by cutting down the weld preparation angle or avoiding the use of surface fillet welds.

Distortion is also much greater if the weld is applied to a region containing compressive residual stress and conversely diminished if the residual stress is tensile. Many bought-in items of stock such as bar and tube will already contain residual stress patterns due to forming, straightening and heat treatment and the interaction of these patterns with subsequent welding operations can be a source of unexpected and frustrating variability in distortion experience. The other important message from this discussion is that symmetrical or 'balanced' welding procedures which are carried out sequentially will not necessarily produce a symmetrical result.

When all possible has been done in terms of design and planning to minimise the effects of distortion, careful inspection of the fabrication during the production process is advised so that developing distortion is monitored and appropriate action taken. Clamping and jigging methods are difficult to assess in a calculated scheme, but are of particular value in the early stages of fabrication where the structure has less in-built stiffness against contraction forces. If accurate final dimensions are required it may in the end be necessary to machine following a full thermal stress-relief.

Appendix

Note 1 The effect of diffusivity

In equation (1.1), Okerblom has neglected a term which takes account of the diffusivity of the material being welded. This neglect could be misleading in certain practical examples, and although for simplicity several calculations in this section have been based on the unmodified equation the following discussion shows how the results can be affected by the diffusivity term.

Wells has given a less approximate formula for the isotherm maximum width as

$$w = \frac{1}{2} \frac{q}{vt} \frac{1}{c\rho} \frac{1}{T} - \frac{4}{5} \frac{\lambda}{v} \tag{1.1a}$$

Repeating equation (1.3) in its corrected form, the longitudinal contraction is given by

$$\varepsilon_l = 0.35 \frac{q}{v} \frac{\alpha}{c\rho} \frac{1}{A} - 0.8 \frac{\lambda}{v} \frac{t}{A} \varepsilon_Y \qquad (1.2a)$$

The in-plane curvature is given correctly by equation (1.4), provided that the modified expression for ε_l has been used, and the correct formulation for transverse contraction is given by equation (1.5).

The correction term will be significant if the diffusivity is high (e.g. in aluminium) or the welding speed is low, especially if the specific heat input is itself small. Using the data in Example 1.1, the correction alters the answer by -3%, although if the problem is repeated assuming aluminium materials, the correction is -7%.

This point may be of interest when experience gained on one material is applied to another. It has been stated for example[1] that stainless-steel fabrications distort more than mild-steel structures, other things being equal, and this is sometimes attributed to the greater expansivity of stainless steel. However, the increased expansion is offset in the factor $\alpha/c\rho$ by a proportionally increased thermal capacity. We may instead look to the lower *conductivity* and hence diffusivity of stainless steel, which will lead to relatively greater distortions at low heat inputs and welding speeds.

Note 2 Welding on thick plates

It will be shown in Chapter 2 that a similar simplified equation for maximum zone width can be found for three-dimensional heat flow. Taking the first term only of that simplification, and following through the previous principles, the longitudinal contraction strain is given by

$$\varepsilon_l = 0.27 \frac{q}{v} \frac{\alpha}{c\rho} \frac{1}{A}$$

Note 3 Recommendations for joint preparations
(Figures 1.42–1.44 and Table 1.4)

Table 1.4

preparations for G.T.A.

Austenitic & Heat-Resisting Steels (adapted from BS 3019 2 1960)

THICKNESS min	max	root face	comments	incl. angle
0·4	0·7	-	*	
0·7	2·0	-	*	
2·0	3·0	-	†	
			NOT RECOMMENDED	
0·7	3·0	-	•	
2·0	3·0	1 5	+	70°
2·0	3·0	-	*	"
3·0	5·0	2·5	* OR dress after welding	
3·0	7·0	1·5	+ OR back-gouge & seal / OR use backing bar	"
7·0	∞	1·5	OR back gouge before welding second side	
7·0	∞	2·0	† OR back gouge & seal r = 4 mm	40°

Aluminium Alloys & Magnesium Alloys (adapted from BS 3019:1 1958)

incl. angle	THICKNESS min	max	root face	comments
	0·9	1·6	-	this prep preferred
	0·9	2·0		
	2·0	4·0		
	2·0	5·0		liable to poor root shape
80°	4·0	6·0	1·5	
	4·0	6·0	1·5	
	5·0	10·0	2·4	back gouge & seal OR use backing bar
	5·0	13·0	2·4	back gouge before welding second side
60° or less if access possible	5·0	∞	3·0	back gouge & seal r = 5 mm

*Inert gas purge on reverse side is beneficial.
†Inert gas purge on reverse side is essential.

70

THICKNESS	JOINT PREPARATION	COMMENTARY
UP TO 8 mm.	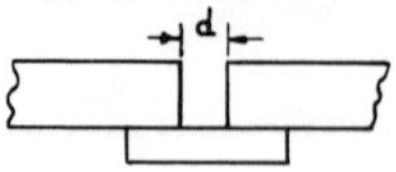	COPPER BACKING BAR SINGLE PASS FROM ONE SIDE d = 0 - 2 mm.
6 - 20 mm.	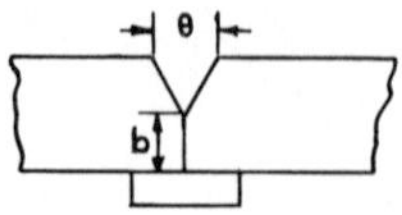	COPPER BACKING BAR SINGLE PASS FROM ONE SIDE θ = 40 - 60° b = 0 - 4 mm.
UP TO 12 mm.	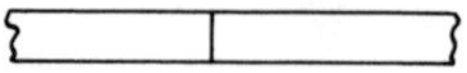	ONE PASS FROM EACH SIDE 2 ELECTRODES IN TANDEM.
10 - 20 mm.	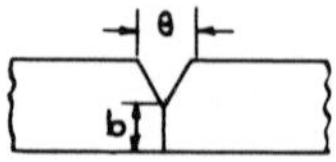	ONE PASS FROM EACH SIDE 2 ELECTRODES IN TANDEM. θ = 65° b = 8 mm.
20 - 40 mm.	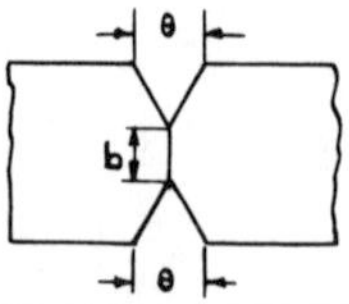	ONE PASS FROM EACH SIDE 2 ELECTRODES IN TANDEM θ = 65° b = 8 mm.

Figure 1.42 Typical joint preparation for submerged-arc welding of steel plate

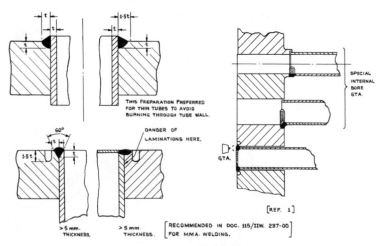

Figure 1.43 Preparations for tube/tubeplate welding

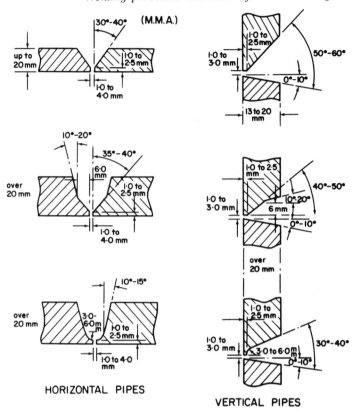

HORIZONTAL PIPES

VERTICAL PIPES

Figure 1.44   Preparations for pipe butt welds

REFERENCES

1. *Control of distortion in welded fabrications*, The Welding Institute, Cambridge
2. HICKS, J. G., *Welded Joint Design*, Granada (1979)
3. OKERBLOM, N. O., *The calculations of deformations of welded metal structures*, 1955 (translated by the Department of Scientific and Industrial Research 1958) HMSO, London
4. ROSENTHAL, D., 'The theory of moving sources of heat, and its applications to metal treatments', *Trans. ASME*, **68**, 849 (1946)
5. RYKALIN, N. N., *Thermal Welding Principles*, I.A.N., SSSR (1947)
6. WICKRAMASINGHE, D. M. G. and GRAY, T. G. F., 'A simple treatment of welding distortion', *Welding Res. Int.* 8, **5**, 409–22 (1978)
7. MASUBUCHI, K., 'Control of distortion and shrinkage in welding', Welding Research Council Bulletin 149
8. BULSON, P. S., *The Stability of Flat Plates*, Chatto & Windus, London

BIBLIOGRAPHY

American Welding Society's *Welding Handbook*, 5th ed., Vols. 2 & 3, Macmillan
MILNER, D. R. and APPS, R. L., *Introduction to welding and brazing*, Pergamon (1968)
PATON, B. E., *Electroslag welding*, American Welding Society (1962)
TYLECOTE, R. F., *The solid phase welding of Metals*, Arnold (1968)
VILL, V. I., *Friction Welding of Metals*, American Welding Society (1962)
Joint preparations for fusion welding of steel, The Welding Institute, Cambridge

Chapter 2

Metallurgical changes and consequences

2.1 Introduction

Not every material in common engineering use can be successfully welded. Some of those even, which are described as 'weldable' in standards and manufacturers' catalogues, require particular processes, and special procedures or treatments which are not convenient in every application. It is reasonable to expect some difficulty, as fusion welding processes in particular subject the material to extremes of heating, cooling, and straining. Also, the chemical composition may be altered, there will be some rearrangement of the constituents, and harmful gases may be absorbed.

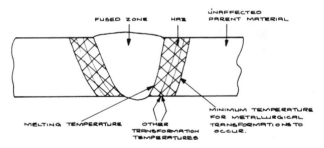

Figure 2.1

It is hardly surprising, therefore, that many metals cannot tolerate such abuse, and as a consequence, cracked, porous, brittle, or weak zones can be formed in the joint. Some improvement in mechanical properties can sometimes be made through post-weld heat treatment, but this is often expensive or impracticable. In some commercial metals, plastic working is used to obtain strength and toughness in the final product. Fusion welding will usually obliterate such beneficial effects, and it will rarely be possible to rework weld zones situated in fabricated structures.

73

The heat-affected zone (see *Figure 2.1*) is in many cases a particularly difficult area. Outside this zone, the original properties are unaffected. Within the melted zone, on the other hand, the metallurgist can add elements to the weld pool via the filler wire, so that the chemical composition is altered, and a more tolerant material may be produced. If it were possible to melt the fused zone without heating the adjacent material above the relevant critical temperature, a large number of metallurgical welding problems would disappear.

2.1.1 Phase diagrams

When the metallurgical aspects of welding are being considered, it is helpful to have some picture of the physical constitution of the material at different temperatures. In particular we usually wish to

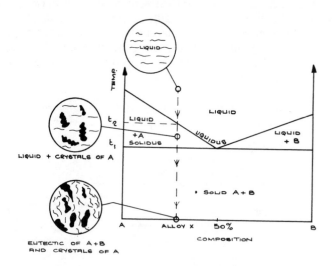

Figure 2.2   Equilibrium phase diagram (eutectic system)

know the state (liquid or solid), and the form and distribution of phases (constituents). Equilibrium phase diagrams are often used to express the relationships between chemical composition, temperature, and constitution. It should be emphasised that such diagrams apply only to equilibrium cooling of the material, in which phase transformations are allowed ample time to take place. They should not be applied directly to treatments such as quenching, tempering, normalising and welding, in which transformations may be suppressed or altered.

Figure 2.2 has been drawn up for a series of imaginary binary (two-element) alloys, consisting of varying proportions of metals A and B. For an alloy X, the cooling metal remains completely liquid

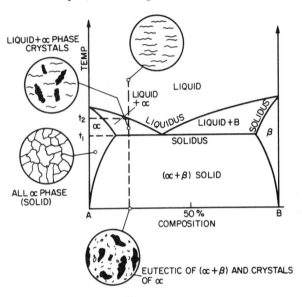

Figure 2.3

to the temperature t_2. Below this temperature, crystals of metal A form within the liquid phase. Below t_1, the alloy is completely solid; the remainder of the liquid transforms to an intimate mixture of A and B, known as a 'eutectic', in which the crystals of A are distributed.

If A and B dissolve each other to a limited extent in the solid state, an alternative diagram is possible as shown in *Figure 2.3*. Alpha (α) phase is produced if B atoms dissolve in A, and Beta (β) phase if A atoms dissolve in B. These phases are normally described as 'solid solutions'. In this figure, a eutectic of $\alpha + \beta$ is produced on solidification.

Phase transformations may also occur in the solid state (below the solidus line in both figures). The transformations may occur by atomic rearrangement or diffusion-controlled reactions. For example, in a 'eutectoid transformation', a single phase transforms to a eutectoid. This happens below the solidus in steels when austenite transforms to pearlite (see Section 2.6).

For a more detailed discussion of phase diagrams see reference 1 (at the end of this chapter).

The aim of the following sections is to explain some of the factors which influence the production of a metallurgically sound joint in common alloys. The discussion concentrates on fusion welding procedures.

2.2　Heat flow—temperature—cooling rate

The Rosenthal/Rykalin equations which were previously discussed in Chapter 1 with reference to distortion, are of equal interest for the assessment of metallurgical response.

In the case of thin plates, equation (1.1a) may be used as a guide to the widths of zones which have been heated to specified temperatures, for example the melting temperature. Repeating the equation here for convenience,

$$w = \frac{1}{2} \frac{q}{vt} \frac{1}{c\rho} \frac{1}{T} - \frac{4}{5} \frac{\lambda}{v} \qquad (1.1a)$$

In many practical situations the first term will dominate showing that, like distortion, the size of fused zones, heat-affected-zones, over-aged zones, etc., all increase with specific energy input.

For moving point source heating of thick plates (see *Figure 1.32*) the relationship is more complicated, although it still depends principally on specific energy input. The basic equations have been used to draw up *Figure 2.4* in which the inner and outer boundaries of a typical HAZ in steel have been plotted for a range of specific heat inputs, and weld travel speeds. Typical heat input rates for a variety of common welding processes are included.

(It is interesting to note that the results in *Figure 2.4* can be simply approximated by an equation:

$$R = \frac{1}{2} \left(\frac{q}{v} \frac{1}{c\rho} \frac{1}{T} \right)^{1/2} - \frac{1}{2} \frac{\lambda}{v}$$

where R is the maximum radius of the isotherm T. The scope of the approximation has not been investigated for other materials. See also reference 14.)

The length-to-width ratio of the isotherms is of some interest, as it is often found that longitudinal cracking of the weld is associated with an elongated isothermal contour as shown for example in *Figure 1.32*. One can imagine that a high transverse thermal stress would be associated with such a pattern. Taking the first term of

(1.1a) together with Rosenthal's equation for the weld centreline temperature

$$T = \frac{q}{2Kt} \sqrt{\left/ \left(\frac{\lambda}{\pi v x} \right) \right.}$$

the length-to-width ratio is given approximately by

$$\frac{x}{w} = \frac{q}{t} \frac{1}{2\pi K T} \tag{2.2}$$

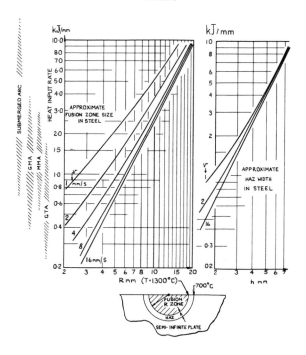

Figure 2.4

Hence a narrow isothermal contour is associated with high heat input rate, and low conductivity.

The cooling rate is also important. *On the weld centreline,* Rosenthal gives

$$\frac{dT}{ds} = -2\pi K c \rho T^3 \left/ \left(\frac{q}{vt} \right)^2 \right. \tag{2.3}$$

for thin plates, and

$$\frac{\mathrm{d}T}{\mathrm{d}s} = -2\pi K T^2 \bigg/ \left(\frac{q}{v}\right) \tag{2.4}$$

for thick plates. Rapid cooling therefore should be associated with *low* heat input rate. A typical centreline cooling curve calculated from equation (2.4), is shown in *Figure 2.5*.

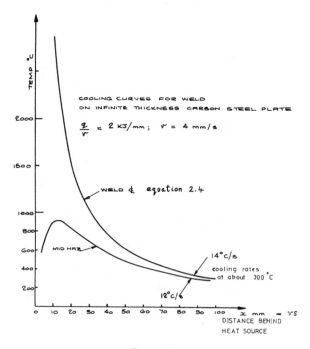

Figure 2.5

The basic heat flow equations have also been used to draw another cooling curve in *Figure 2.5* for a point in the HAZ which reaches 900 °C. It is clear that a large distance behind the heat source, the centreline and HAZ cooling rates become very similar, and there would be little error therefore in using equation (2.4) for the HAZ. At closer distances, however, the cooling rates are vastly different.

In certain steels where hydrogen-induced cracking is a risk, it is particularly important to slow the HAZ cooling rate through a critical temperature band around 300 °C. A useful improvement can be made by preheating the base plate, whereupon the temperature T figuring in all the previous equations becomes $(T_{actual} - T_{base})$ without any alteration in the isothermal pattern. Hence, for example, if a preheat of 100 °C is used, the cooling rate will be reduced in the ratio $[(300-100)/300]^3$ for thin plates and $[300-100/300]^2$ for thick plates. Clearly, preheat has less effect on cooling rate at higher temperatures.

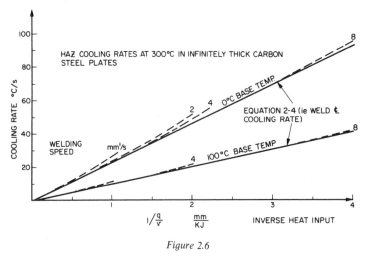

Figure 2.6

The cooling rates at 300 °C, with and without preheat, are further illustrated in *Figure 2.6* for different heat input rates. (Solid lines identify centreline cooling rates, and the broken lines indicate small corrections necessary for the heat-affected-zone position.) Hence, in materials which react badly to quenching (high-carbon steels) the very welding conditions which reduce cooling rate (high specific energy input), will lead to a large weld and heat-affected zone size. On the other hand, low specific energy input processes will be very suitable for materials which suffer if exposed to high temperatures for long periods (e.g. heat-treatable aluminium alloys).

2.3 Weld chemistry—dilution

The metallurgical response of each part of the weld joint will depend greatly on its chemical composition. In fused zones, one finds that for most processes the materials which are melted into the weld pool become well mixed by stirring arc forces and convection currents.

Consequently the final fusion zone composition is relatively uniform, and depends only on the proportions and initial compositions of the constituents. The proportion is often expressed as a dilution ratio, defined as the fraction of parent material dissolved in the final fused zone size. Thus for an autogenous weld (no filler) dilution is

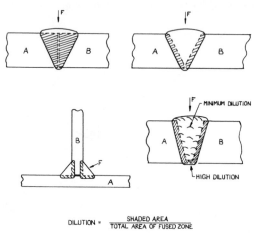

$$\text{DILUTION} = \frac{\text{SHADED AREA}}{\text{TOTAL AREA OF FUSED ZONE}}$$

Figure 2.7

technically 100%, whereas for a minimum penetration weld, dilution will approach zero. If the initial joint configurations and final fused zone shapes are known, as in *Figure 2.7*, the dilution may be estimated graphically. This, in turn, establishes the fused zone compositions, namely

$$W = kF + D_A(A - kF) + D_B(B - kF) \qquad (2.5)$$

where W is the final percentage of given element in the fused zone, F the initial percentage of given element in filler material, k the proportion of filler element which transfers across the arc into the weld pool, D_A the dilution fraction of parent metal A, D_B the dilution fraction of parent metal B, A the initial percentage of given element in metal A, B the initial percentage of given element in metal B.

It can be readily appreciated from *Figure 2.7* that the proportion of parent metal which is melted depends on the edge preparation and the welding process parameters. Thus, high dilution will occur for single-pass square-close-butt welds, especially where large specific energy input processes are used (e.g. submerged arc). *Table 2.1* gives some idea of the variations in dilution which can occur. Similarly, dilution may vary within a multipass weld, the maximum value probably occurring in the root area, with some reduction close to the fusion line, reducing possibly to zero for centre weld zones near the surface.

Table 2.1 TYPICAL DILUTION RATIOS FOR GTA WELDS (IN ALUMINIUM)

Thickness (mm)	Joint preparation	Dilution (%)
3	square close butt	65–100
10	square close butt, single pass	85–90
10	square close butt, two pass (both sides)	75–80
10	single V, single pass	55–60
any thickness	unprepared fillet weld	30–40

The practical implications need to be appreciated when edge preparations, filler materials and welding processes or parameters are being considered. For example, it will be shown later that solidification cracking can arise in certain steels if the sulphur and phosphorus levels in the fused zone are too high. These critical levels may be easily exceeded in commercial quality plate and forgings. Cracking is therefore a definite risk if a high-dilution process is used. Nevertheless, high-dilution procedures are used in practice, perhaps because they are technically convenient, as in the case of autogenous tube-to-tubeplate welds[2], or because they have economic attractions as in the case of square-close-butt welds which are made using the submerged-arc process.

Dilution must also be carefully estimated for dissimilar metal joints where undesirable compounds would result from a straight mix of the two components. If a suitable filler metal composition is chosen on the basis of equation (2.5), a metallurgically satisfactory transition can be made.

The composition of backing strips is, of course, of equal importance, and cracking faults have frequently been traced to the use of unspecified temporary backing material which was 'just lying around handy'. Also, if copper forming bars etc. are in contact with a steel weld, care must be taken that the steel is not embrittled through copper absorption.

Finally the mechanical properties of the fused zone will, of course, depend on chemical composition (even in the absence of cracking). However, unless special strengthening treatments have previously been applied to the parent material (e.g. quench and tempering), the alloy level in the weld does not usually need to be as high as in the parent material to achieve the same yield strength. This is attributed to the extensive plastic deformation suffered during the weld cycle, which strengthens the matrix. In steels, a normalising treatment (see page 107) reduces the yield strength to the level expected of wrought materials which have the same composition.

2.4 Gas absorption—porosity

It was emphasised in Chapter 1 that it is vital to exclude air from the molten weld pool. There are of course other sources of harmful gas— for example, moisture, oils, or corrosion products. The parent material itself may also contain absorbed gas which will be released on melting.

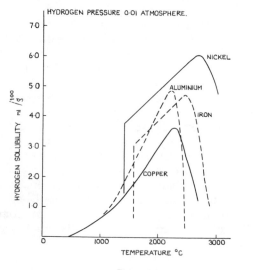

HYDROGEN SOLUBILITY-TEMPERATURE CURVES (AFTER HOWDEN & MILNER. BRITISH WELDING JOURNAL. V.IO N°6 JUNE,1963.)

Figure 2.8

The gases which most often cause trouble are hydrogen, nitrogen, and oxygen. Typical faults include porosity and piping, cracking, and clusters of oxide inclusions.

The root of the problem is that gases are more easily dissolved in the molten metal at high temperature, as shown in *Figure 2.8*. They may subsequently become trapped in the solid metal if cooling is rapid. The gas may either be retained in the microstructure or may form bubbles which can become trapped as porosity in the fast-freezing metal. One cannot say that gas retained in the weld in whatever form is invariably harmful; specific problems will be noted later. However, the leakage of pressure parts can often be traced to linked porosity bubbles.

In some cases, gas bubbles may emerge from pockets within the parent material; for example, from pores in castings or plate lamination defects which are cut open by the weld edge preparation (see *Figure 2.9*). Poor design of the joint configuration can also trap air which is subsequently forced to expand through the weld pool[2].

Porosity formation depends on three consecutive factors—the time allowed for absorption at high temperatures, the time required

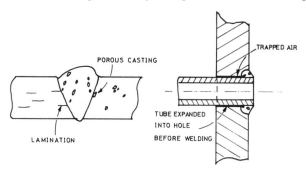

Figure 2.9

for bubble nucleation and growth, and the ability of the bubbles to rise through the weld pool and break free at the weld surface. For these reasons, alterations in the process parameters might appear to have contradictory effects. In aluminium and steel for example, raising the welding speed (which increases the cooling rate) usually promotes porosity, as the generated bubbles are more easily trapped. In titanium on the other hand, an increase in welding speed can *reduce* porosity. (This may mean that hydrogen is retained in the microstructure.)

If low-heat-input, low-hydrogen manual-metal-arc practice is specified, starting porosity may be troublesome, and occurs because the small quantities of shielding gas evolved from the electrode coating may be insufficient to exclude air until the process is properly established.

2.4.1 Hydrogen

Porosity due to hydrogen is typically associated with the welding of aluminium, copper, titanium, magnesium and niobium. Moisture and organic compounds form the main sources. In steel welds, porosity is less troublesome than cracking, which occurs because excess hydrogen diffuses to the heat-affected zones where the microstructure tends to be brittle.

2.4.2 Nitrogen

Air is the main source of nitrogen contamination. Porosity seems to be particularly associated with the gas-metal arc welding of carbon steel, stainless steel, and nickel. Usually, one finds that the gas shield has been contaminated either at the source of supply, or owing to air entrainment at the shielding nozzle. It is important to shelter the welding point from cross draughts, and to make sure that the gas flow rate is adequate and nonturbulent. The joint configuration can often be designed to shelter the weld pool and confine the gas supply; for example, a weld in a U groove is usually much easier to shield than a fillet weld.

2.4.3 Oxygen

Porosity formation due to oxygen absorption is normally indirect, in that troublesome gas is usually a reaction product (for example, carbon monoxide—formed between oxygen and carbon). In poorer quality 'semikilled and unkilled' steels, oxygen is also released by the parent material. Excess oxygen is usually mopped up by the addition of traditional steelmaking deoxidants to the filler material—that is, aluminium, silicon, manganese, and titanium. As a result, inclusions are formed. Unfortunately, excess unused deoxidant often reduces the notch toughness of the weld deposit. Remember also that the deoxidant level may be insufficient, if too much *dilution* takes place.

In conclusion, *Table 2.2* gives a convenient guide to the most common sources of porosity in various materials. Apart from the avoidance of air pockets as in *Figure 2.9*, control of porosity reduces to the correct selection of the parent/filler metal and welding process combination, together with clean, dry welding practice.

2.5 Cracking

Cracks can arise in various parts of the joint, and at various stages of the weld cycle and fabrication programme. Although the causes may vary in detail, the broad reason for cracking is that the material has insufficient strength or ductility at the relevant stage to tolerate

Table 2.2 CAUSE AND PREVENTION OF POROSITY

Metal	Absorbed gas causing porosity	Chemical reaction in weld pool	Prevention typical deoxidant
Mild steels and low-alloy steels	oxygen and	$C + [O] = CO_{(gas)}$	Ti, Al, Si, Mn
Stainless steels	nitrogen		Ti, Si
Nickel†			Ti
Copper and copper alloys	hydrogen oxygen nitrogen	$[2H] + [O] = H_2O_{(gas)}$	P Ti B
Aluminium†	hydrogen		
Titanium†	hydrogen oxygen argon		
Magnesium†	hydrogen		
Niobium†	hydrogen		

†Pure shielding gas important.

the stresses and strains imposed by the transient thermal strain field or the subsequent residual stresses.

In chronological order, the main cracking mechanisms to be discussed are solidification cracking in the weld metal, burning and hydrogen-induced cracking in the HAZ, and reheat cracking which arises in multirun welding or during subsequent heat treatment (e.g. stress relief). Lamellar tearing is also mentioned.

2.5.1 Solidification cracking

The molten weld pool freezes progressively from the fusion line towards the centre which, being at the highest temperature, solidifies last. Concentrations of alloying elements and impurities are pushed ahead of the inward growing crystals or dendrites. The last stages of solidification, where dendrites interlock, are critical because low-freezing point liquid films separating newly formed crystals produce a region of weakness and cracking occurs if transverse thermal strains are high. Low-freezing point liquid films are either caused by impurities such as sulphur and phosphorus in steels or by alloying elements which create a wide freezing range such as magnesium in aluminium/magnesium alloys. Any cracks which are formed may be healed if liquid metal can flow freely into the crack; but eventually the supply of liquid will be shut off by the interlocking dendrites. *Figure 2.10a* illustrates solidification cracking typical of aluminium alloys.

The end of the weld run is also a typical location for solidification cracking. If the heat source is cut off abruptly, the material cools rapidly, leaving a longitudinal or star-shaped centre crack. If the arc

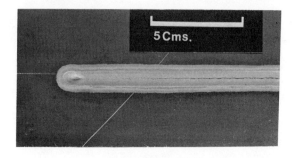

Figure 2.10a   Solidification cracking

Figure 2.10b

forces are high. the weld pool freezes with a concave or 'cratered' shape. For this reason, special techniques are used to reduce heat input gradually, or else the weld end is made on a finishing block which is subsequently cut off. (Remember that the composition of the finishing block is important from the point of view of dilution at the weld end.)

Solidification cracking can be controlled through weld metal composition, process parameters, and joint design/edge preparation; probably in that order of significance. Weld metal compositions which have a wide freezing range should be avoided; and impurities which promote the formation of low freezing point liquid films must be strictly controlled. Process parameters which stimulate coarse dendritic growth (generally those which slow the heat abstraction rate) can lead to cracking, as impurities are then more likely

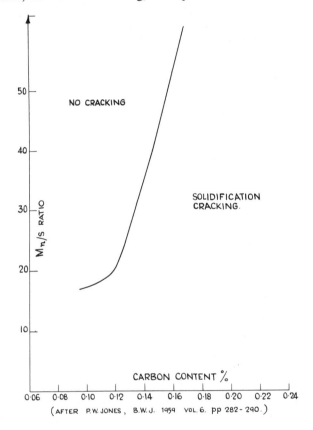

(AFTER P.W. JONES, B.W.J. 1959 VOL.6. pp 282-290.)

Figure 2.11

to segregate towards the weld centreline. The relationship between cracking susceptibility and weld travel speed is unfortunately complex for most materials—in the case of gas-tungsten-arc welding of low-alloy steel sheet, it has been reported that an increase of speed increases the cracking tendency at first and reduces it at higher speeds. Joint preparations or welding procedures which force the fusion zone into a narrow deeply penetrating shape cause trouble

(*Figure 2.10b*). As in other forms of cracking, a joint configuration which maintains restraint or forms a geometric stress concentration in the weld should be avoided. Root runs are particularly liable to cracking for this reason.

In steels, impurity films forming at the grain boundaries are the main culprits. Sulphur, phosphorus, boron, copper and arsenic are the most troublesome elements. Sulphur, for example, promotes the formation of low-freezing point sulphide films. (It is worth noting

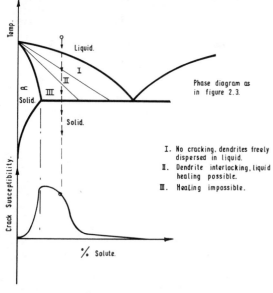

(After J.C. Borland. B.W.J. 1960 Vol. 7 p.508.)

Figure 2.12

that sulphur is an important constituent of free-cutting steel, which is widely used for machine parts.) The impurity level in the weld pool will depend on the initial impurity content of the steel and the dilution ratio but, as previously noted, impurities are liable to concentrate in the centre of the weld. Sulphur also has an unfortunate tendency to migrate to particular parts of the workpiece where welds may be applied (for example, at the surface of tube plate forgings), and this may upset the application of autogenous tube-to-tubeplate welds of the type shown in the Appendix to Chapter 1 (see also reference 2).

Of course, one can avoid the problem by specifying high-purity steel, but this may be an expensive solution. The alternative is to tie up the harmful element in another form which is less damaging. For example, manganese additions to the filler wire can be used to bind the sulphur in the form of manganese sulphide inclusions. However, the benefits are less effective in higher carbon and alloy steels as shown in *Figure 2.11*.

In aluminium alloys, crack susceptibility can often be explained via the phase diagram; for example, as shown in *Figure 2.12*. If there is a large difference between the liquidus and solidus temperatures, i.e. a wide freezing range, there will also be a large relative contraction strain in the intermediate zone, and the supply of liquid metal for crack healing will be difficult to maintain.

For such alloys, therefore, it is common practice to specify a composition which is either close to the pure metal or well above the critical range. For example, in aluminium–magnesium alloys the possibility of solidification cracking is greatest when the magnesium content of the weld metal is less than 3% and consequently a 5% magnesium filler wire composition is used so that, with due allowance for dilution, the weld metal has more than 3% magnesium.

2.5.2 Heat-affected-zone cracking

Burning

Although parent material in the HAZ does not melt as a whole, the temperature may be so high close to the fusion line that constituents having lower melting points than the surrounding matrix become liquid. Fine cracks or tears may result if the thermal stress is high enough. Although the cracks are small, they will be situated close to the weld toe which is a point of stress concentration (see Chapter 3 and Section 4.3), and therefore they can be readily extended by fabrication stresses, or in service. Such cracks have been identified occasionally as the initiation sites for fillet weld fatigue failures.

This phenomenon called 'burning' or 'liquation cracking', is found in ferritic and austenitic steels and in some aluminium/copper and aluminium/magnesium alloys. In steels, the low-melting point films may be formed from impurities such as sulphur and phosphorus, boron, arsenic and tin. As in the case of weld metal solidification cracking, increased carbon, sulphur and phosphorus levels make the problem more acute and increased manganese improves the situation. Steelmaking practice is as important as composition—basic electric steels being more liable to cracking than open-hearth steels. In the case of aluminium alloys, cracking is due to lower-melting point films rich in copper or magnesium.

Liquation cracking occurs in the HAZ and therefore cannot be directly controlled via additions to the weld pool (except that a low-yield point weld metal reduces the thermal stress on the HAZ). Low sulphur and phosphorus levels are considered essential in high-yield or high creep strength steels, but otherwise it is difficult to give hard-and-fast rules on burning susceptibility. As far as welding process variables are concerned, one should aim at reducing the hold time at the liquation temperature.

Hydrogen-induced cracking

This form of cracking which occurs in the heat-affected zone at temperatures less than 200 °C is a major source of trouble in carbon and low-alloy steels. Cracks may form within minutes or may be

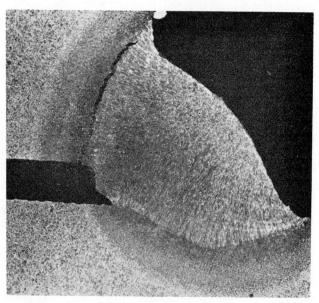

Figure 2.13   Hydrogen-induced crack

delayed for several days after welding. This could, of course, be a considerable embarrassment to a fabrication schedule, particularly as the cracks themselves are difficult to detect nondestructively (see *Figure 2.13*).

As discussed in Section 2.4, the weld metal may be supersaturated with hydrogen, which is driven out of solution on cooling and diffuses to the HAZ which will usually have a greater absorption capacity. If the microstructure there is susceptible to embrittlement, and sufficient residual stress is present, cracking will occur. The important variables are shown in *Figure 2.14*.

The susceptibility of the HAZ microstructure to cracking arises through the formation of brittle phases. Martensite, particularly in the twinned form, is extremely brittle, whereas microstructures consisting of ferrite and pearlite are more resistant. Martensite formation is encouraged by high cooling rates and a chemical composition which is high in carbon and alloying elements such as manganese, chromium, nickel, silicon and molybdenum.

The factors controlling cooling rate have already been outlined in Section 2.2 but some of the practical implications of preheating require elaboration. Despite the description '*pre*heating' the desired base temperature must be maintained throughout welding and for some time after completion, if the benefits are to be realised. Furthermore, a sufficiently wide band of parent material must be heated, and care must be taken that the heating pattern does not itself generate large thermal stresses elsewhere in the structure. Many heating methods are used—gas burners, and electrical resistance or induction element arrays. A high workpiece temperature will also increase welder discomfort and add a further delay to work schedules. For these reasons, preheating costs may constitute a large proportion of the total fabrication price. For certain structures (much of ship hull construction) it is in any case impracticable. Apart from the effect on cooling rate, high base temperatures will encourage hydrogen diffusion out of the weld zone, and even a moderate preheat (say 50 °C), will serve to drive off moisture from the workpiece.

'Carbon equivalent' formulae are widely used as empirical guides relating composition and cracking tendency. For example an early formula due to Dearden and O'Neill[3] gives

$$C_{equiv} = \%C + \frac{\%Mn}{6} + \frac{(\%Cu + \%Ni)}{15} + \frac{(\%Cr + \%V + \%Mo)}{5}$$

These formulae greatly oversimplify the true behaviour. Furthermore, they are not always drawn up on the basis of real cracking experience, but may be determined by examination of peak hardness in the HAZ, metallographic examination for martensite, or bend ductility. These other phenomena are easier to measure in the laboratory but are not necessarily related to cracking. In early work on structural steels it was found that if the hardness was controlled below 350 HV cracking was prevented. However, it has become clear that hardness can exceed this level without cracking if hydrogen

92

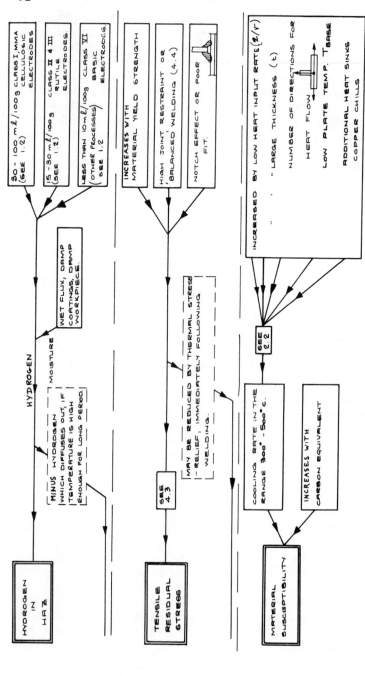

Figure 2.14 Factors influencing hydrogen-induced cracking

levels or residual stresses are low. Again for other steels, there is no reason why a particular hardness value should be a safe guide, as the link between hardness and cracking is tenuous.

Even when the formula is based on laboratory cracking tests, factors other than composition may not be controlled to a standard pattern (e.g. hydrogen content, fit-up, restraint, cooling rate, parent material grain size, etc.). It is not surprising, therefore, that the different formulae attribute different influences to the various elements. (In other words the denominators vary from one formula to another.) For example, in a cracking control scheme due to Winterton shown in *Table 2.3*, a *negative* influence on cracking is attributed to Mo and V.

Table 2.3 WELDING PRECAUTIONS FOR LOW-ALLOY STEELS
(after Winterton, K., *Welding Journal*, **40**, 253s–258s, 1961)

Carbon equivalent	Welding procedure
less than 0.40	No precautions, cellulosic or rutile electrodes may be used
0.40–0.48	Use: ordinary electrodes & low preheat (100–200 °C) OR low-hydrogen (basic) electrodes
0.48–0.55	Use: ordinary electrodes and 200–350 °C preheat, OR austenitic electrodes (see p. 113) OR GMA process
greater than 0.55	Use: low-hydrogen electrodes and 200–350 °C preheat OR austenitic electrodes OR GMA process

$$\text{Based on } C_{equ.} = \%C + \frac{\%Mn}{6} + \frac{\%Ni}{20} + \frac{\%Cr}{10} - \frac{\%Mo}{50} - \frac{\%V}{10} + \frac{\%Cu}{40}$$

Winterton's formula is based on an examination of a variety of carbon equivalent formulae for low-alloy steels. These other formulae covered steels with compositions as follows: 0.12–0.30% C, 0.25–1.62% Mn, 0.04–0.21% Si, 0–5.4% Ni, 0–1.7% Cr, 0–0.64% Mo, 0–0.14% V. Welders would no doubt be horrified at the prospect of using 350 °C preheats (see 0.48–0.55 carbon equivalent line), particularly as a more practical alternative suggests itself, namely, the use of low-hydrogen electrodes with 100–200 °C preheat. Other crack-susceptibility formulae directly include the effect of the hydrogen potential of the welding process and plate thickness[4] and the effect of restraint[5]. Dolby[15] draws attention to the superior weldability of low-carbon, high-strength structural steels, demonstrated in a $C_{equ.}$ formula which attributes a factor of 1/20 to Mn and Cr.

The message is quite clear, therefore, that it could be very misleading to use a given formula in isolation. It must be seen as an essential part of a particular cracking control scheme which has been found to be effective in defined circumstances and for a specified range of steel compositions.

Among the many laboratory cracking tests which have been devised, the CTS (controlled–thermal–severity) test has achieved wide acceptance, as it incorporates most of the known variables with the exception of large-scale structural restraint. The test specimen form is shown in *Figure 2.15*, and recommended testing procedures are described in reference 6.

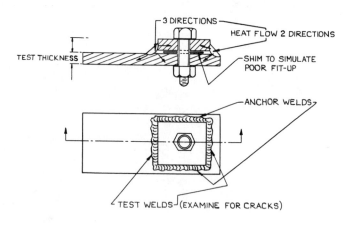

Figure 2.15   CTS specimen form

One of the most comprehensive cracking control schemes for structural steels has been produced by Bailey[7]. The interrelated effects of cooling rate, material composition and hydrogen level are displayed in a commendably simple diagram which is reproduced here as *Figure 2.16a*, with *Figure 2.16b* given as a key. The material susceptibility is measured for BS 4360 steels in terms of Dearden and O'Neill's carbon equivalent formula (bottom right-hand side of *Figure 2.16b*). It is worth noting that the chemical analysis required by BS 4360 is made on molten steel in the ladle. However, the purchaser has the option to specify the final product analysis, which may well be different.

Common fusion welding processes are classified in three groups (A, B, C in the bottom right-hand corner of *Figure 2.16b*) according to their hydrogen-producing potential.

95

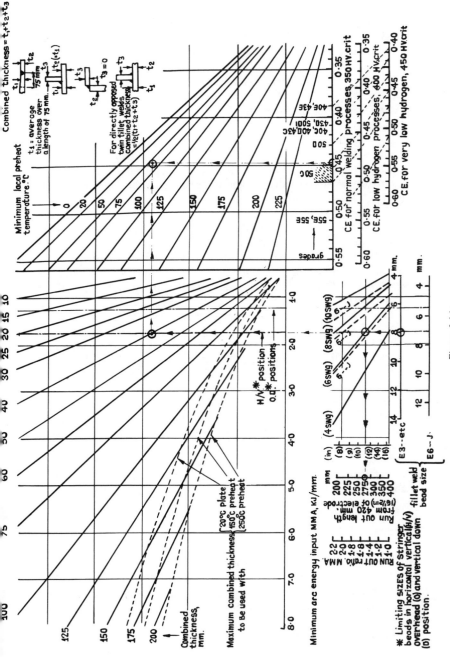

Figure 2.16a

A. Rutile and cellulosic electrodes (class I, II, III) or improperly dried, low-hydrogen electrodes (class VI)
B. Poor-quality CO_2 welding (dirty wire or workpiece) or low-temperature baked class VI electrodes
C. High-quality GMA, GTA, submerged-arc with dry flux, or thoroughly baked class VI electrodes

Empirical critical hardness levels are shown for the three classes of processes.

The cooling rate depends on the preheat temperature (top right-hand side of *Figure 2.16a*), the thermal capacity of the joint configuration (total plate thickness presented to the heat source— as in top left-hand side of *Figure 2.16a*), and the heat input rate (bottom left-hand side of *Figure 2.16a*).

A number of alternatives are given for determination of heat input rate. If the welding conditions are known or specified, a direct determination of q/v can be made in kJ/mm. Otherwise, the rate may be related to known MMA procedures, as in the 'indirect' route at the extreme left of *Figure 2.16b*. These procedures may be specified in terms of electrode size and 'run-out' ratio (length of electrode used/weld length) or in terms of fillet size. In the latter case, a small adjustment is made for the electrode type, as class III electrodes deposit more metal for a given heat input than class VI. The maximum heat inputs which can be employed in the horizontal/ vertical, overhead, and vertical down positions are also indicated as vertical lines drawn at 1.8 and 1.2 kJ/mm. A similar relationship between fillet size and arc energy for submerged-arc welds is given in reference 8.

Bailey's procedure assumes poor fit-up. Observance of the conditions shown will not guarantee freedom from cracking in every case, but will greatly reduce the risk of serious cracking.

Note: the diagram is only valid for C–Mn steels; indeed it is unlikely that a similarly simple relationship between cooling rate and susceptibility could be found for more complicated alloys.

Example 2.1

A beam of I section is to be fabricated from 13 mm thick plate with 8 mm leg-length fillet welds between the web and flanges. The specified steel corresponds to a carbon equivalent of 0.45 using Dearden and O'Neill's formula. Determine suitable welding procedures.

For the web-to-bottom flange welds, which are carried out in the

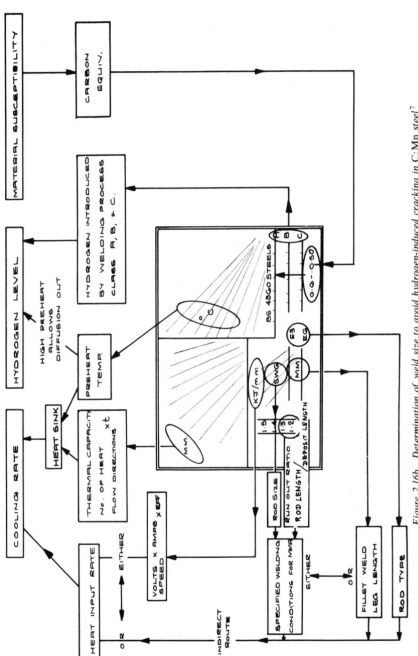

Figure 2.16b Determination of weld size to avoid hydrogen-induced cracking in C:Mn steel[7]

horizontal/vertical position (see *Figure 1.2*), the heat input cannot exceed 1.8 kJ/mm (denoted by vertical line on *Figure 2.16a*). This might correspond in practice to welding conditions of 220 A, 24 V, applied to a 6 s.w.g. MMA rutile electrode, which is run out to a length of 275 mm to give the desired 8 mm fillet. For a combined thickness of 40 mm, a preheat of about 60 °C is indicated (see arrows).

If preheat is inconvenient, there are a number of alternatives:

1 The material specification can be altered to give a maximum carbon equivalent of about 0.42.

2 The hydrogen potential of the welding process could be reduced by other classifications (groups B or C). If class VI electrodes are used there may be some difficulty in producing the required weld size in one run.

3 The twin fillet welds could be carried out simultaneously (arcs directly opposed), thus doubling the heat input rate (or effectively halving the combined thickness).

4 If the specified fillet size is increased and the beam is tilted to produce a down-hand welding position, the heat input rate may be increased to over 2 kJ/mm.

If the web-to-upper-flange welds must be carried out in the overhead position, the heat input rate will be restricted further. In this case, a preheat of over 125 °C is necessary, or very low hydrogen practice must be specified. Also the required fillet size cannot be made in one run.

2.5.3 Lamellar tearing

This form of cracking has already been illustrated in *Figure 1.25*. Tears normally occur in the parent material close to the HAZ outer boundary, in joints where the fusion line is parallel to the plate rolling plane. Understandably the risk appears to be greater when the weld is also parallel to the rolling *direction*. Cases have been reported in a variety of plain carbon, carbon–manganese, and low-alloy steels. Laminar clusters of nonmetallic inclusions are at the root of the trouble, and steels which have low inclusion contents (achieved perhaps by vacuum degassing) give significantly less trouble. Sadly, according to present experience, hardly any grades seem to be positively trouble-free.

Accepting that there may be a poor ductility in the thickness direction in many batches of plate, there are however certain welding and design features which contribute to tearing: for example, the joint configuration may impose a high restraint on contractions normal to the incipient cracking plane, and there is evidence[9] that the transverse ductility of certain susceptible materials experiences a temporary drop in the 200–400 °C range, which partly explains the location of typical cracks just outside the HAZ. Apart from changes in the joint design in the spirit of *Figure 1.25*, transverse restraint can be minimised by intelligent choice of welding sequence, for example by making the difficult joint first.

The effect of heat input rate is not as straightforward as it was in the case of hydrogen-induced cracking. This may be expected, since increasing heat input can produce counteracting effects on the temperature and the transverse stress in the troublesome zone. In the cracking tests reported in reference 9, increased heat input rate *reduced* the cracking tendency—whether this was related to local transient thermal stress patterns or critical temperature range effects is not clear. However, remembering that contraction is proportional to q/v, in a restrained joint the contraction stresses will increase approximately with heat input. It might therefore be suggested that the welding conditions which minimise distortion (excluding balanced welding however) will also minimise tearing in rigid configurations.

Despite the best precautions, lamellar tearing has an unfortunate tendency to arise in production when there is little opportunity to alter materials or design. As a makeshift solution, it may help to change to a lower yield strength weld metal, or to cut out the offending zone, and 'butter' with a softer layer.

2.5.4 Reheat cracking

Removal or reduction of residual stresses after welding by thermal stress relief is recommended for many structures. In principle, the structure is heated to a temperature which allows rapid creep (about a third or half the melting point) and held at that temperature until the elastic stresses are relaxed. To reach the desired temperature, the structure may have to be heated through a range in which the HAZ microstructure has very poor creep ductility. In fact the stress-relieving temperature may well be in the middle of such a range. The possibility arises that creep cracking may occur (usually in the HAZ) before the residual stresses have relaxed appreciably.

The residual stress which drives the process may be supplemented by transient tensile thermal stresses in the weld zone resulting from rapid nonuniform heating in the furnace. Geometric stress raisers (e.g. fillet toes) and prior cracks (e.g. liquation cracks) accentuate the problem. Reheat cracking failures have been experienced in nickel-based alloys, austenitic stainless steels, creep-resisting ferritic steels and in certain precipitation hardening high-yield steels. A typical intergranular crack is shown in *Figure 2.17*.

Creep deformation takes place within the grains, and by grain boundary sliding. If the grains are strengthened relative to the grain boundaries—perhaps by precipitation of alloy carbides and

Fillet toe

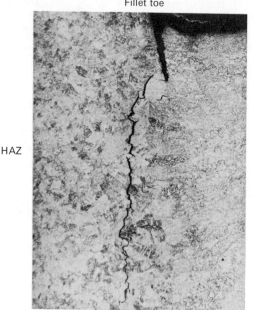

HAZ   Weld metal

Figure 2.17   Reheat cracking in a low-alloy steel
(*Courtesy Babcock and Wilcox (Operations) Ltd.*)

nitrides—premature creep rupture becomes possible. For example, vanadium carbide precipitates are responsible for the reheat cracking of Cr–Mo–V steels.

Creep processes are diffusion controlled, and are therefore sensitive to temperature and holding time. This is demonstrated in *Figure 2.18* for a number of materials which are known to be susceptible. This figure suggests that, for some materials, trouble may be avoided by applying a rapid heating rate through the critical temperature range (without, of course, introducing tensile thermal stress). This approach has been used successfully for austenitic steels and certain age-hardening nickel-based alloys. However, for materials such as the $\frac{1}{2}$Cr–Mo–V steel shown in *Figure 2.18*, the critical time is too short for such a technique.

It is obviously difficult to avoid reheat cracking if the material is susceptible. Of course one should try to avoid the coincidence of

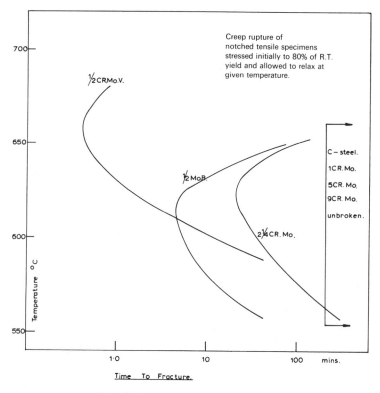

Figure 2.18

geometric stress raisers and HAZs, and in this respect careful weld shaping and dressing may help. Control of heating rate has been mentioned, and it would seem technically possible by judicious use of *non*uniform heating, to generate a transient compressive stress in the weld zone until the troublesome temperature band is passed. (The area which is at the highest temperature will be in compression.)

Nevertheless, it is clear that the problem is best controlled by specifying nonsusceptible materials. Nakamura and others[10] have attempted to define susceptible compositions in low-alloy steels containing up to 1.5% Cr. They anticipate trouble if the summation $\sum Rh = \%Cr + 3.3(\%Mo) + 8.1(\%V) - 2$ exceeds zero. Although this equation underestimates the detrimental effect of boron in the molybdenum-bearing steels, *Table 2.4* shows that it is a useful guide

Table 2.4 SUSCEPTIBILITY TO REHEAT CRACKING OF VARIOUS LOW-ALLOY STEELS

Steel composition	$\sum Rh$	*Actual behaviour*
AB212BQ (0.26% C–0.70% Mn)	−2	No cracking
A533A	$-\frac{1}{2}$	No cracking
A533B	$-\frac{1}{4}$	Susceptible
A387BQ	0.60	Cracked
A542 (0.11% C–0.4% Mn–2.2% Cr–0.96% Mo)	3.4	Cracked
A543	+1.0	No cracking
A517A	−1.0	No cracking
A517B (0.17% C–0.9% Mn–0.5% Cr–0.20% Mo–0.04%V)	−0.60	Some cracking
A517E (0.15% C–0.70% Mn–1.76% Cr–0.50% Mo–0.07% TiB)	1.40	Cracked
A517F (0.18%C–0.9%Mn–0.9%Ni–0.5%Cr–0.5%Mo)	0.50–0.90	Bad cracking
A517J	−0.06	Borderline
Mo/B–$\frac{1}{2}$% Mo	−0.30	Not susceptible
Mo/B–1% Mo	1.30	Susceptible
Mo/B–1$\frac{1}{2}$% Mo	3.00	Susceptible
$\frac{1}{2}$Mo/B–0.01% V	−0.04	Not susceptible
$\frac{1}{2}$Mo/B–0.04% V	0.20	Susceptible
$\frac{1}{2}$Mo/B–0.30% V	2.50	Susceptible
C–Mn	−2.0	Not susceptible
C–Mn–0.002% B	−2.0	Not susceptible
C–Mn steels (B.S. 1501, B.S. 968)	−2.0	Not susceptible
$\frac{1}{2}$Mo–B	−0.50	Susceptible
$\frac{1}{2}$Cr–Mo–V	2.50	Susceptible
1 Cr–Mo	0.50	Not susceptible
2$\frac{1}{4}$ Cr–Mo	3.40	Borderline
5 Cr–Mo	4.50	Not susceptible
9 Cr–Mo	9.70	Not susceptible
C–Mn–0.2% Mo–0.04% V	−1.00	Cracked
C–Mn–0.25% Mo–0.15% V	0.10	Cracked
$\frac{1}{2}$Cr–$\frac{1}{2}$Mo–V	>0.50	Cracked
1 Cr–1 Mo–0.35% V	5.20	Cracked
1.7% Cr–0.2% Mo–0.28% V	2.60	Cracked

After Nicholls. R. W.. *Welding in the World*. Vol. 7. No. 4 (1969).

to cracking tendencies. Once again it is dangerous to extrapolate this formula outside the range of steels examined (the 2$\frac{1}{4}$–9%Cr steels shown are resistant to reheat cracking despite the high Rh index).

2.6 Particular materials—weldable specifications—steels

2.6.1 The iron/carbon system

It is difficult to understand the welding response of steels without some acquaintance with the variety of metallurgical phases which can be formed. Some of these (martensite, pearlite, etc.) have already been mentioned, and although a brief account is given here, the reader is advised to obtain fuller descriptions from a standard metallurgical textbook such as reference 1.

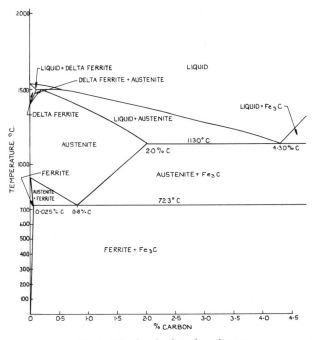

Figure 2.19   Iron/carbon phase diagram

The iron/carbon phase diagram (*Figure 2.19*) seems extremely complex at first sight. However, attention may be confined to a relatively small area of the diagram, as weldable steels typically have carbon contents less than 0.5%.

At high temperatures, *austenite* is formed. It has a face-centred cubic (fcc) crystal structure which dissolves up to 2% carbon, and is soft and ductile. Under equilibrium cooling conditions in the composition range of interest, it transforms to ferrite and pearlite. *Ferrite* has a body-centred-cubic (bcc) structure which can hold only 0.025%

carbon (see *Figure 2.20*). It is also soft and ductile. *Cementite* is the name given to iron carbide (Fe_3C) which is hard and brittle. *Pearlite* is composed of alternate platelet layers of ferrite and cementite, and has properties somewhere between these two phases.

The conversion of austenite to pearlite is relatively slow, as the crystal structure has to change (fcc/bcc), and the surplus carbon needs time to diffuse to the cementite plates. If insufficient time is

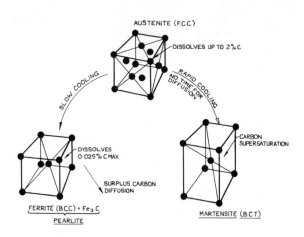

Figure 2.20

allowed, the austenite may transform to a variety of other structures depending on the chemical composition and the cooling rate. *Fine pearlite* may be formed. This phase is much harder than pearlite formed during slow cooling, and the plate widths are much smaller. *Martensite* is formed if the cooling rate is so high that the transformation of austenite to pearlite is suppressed. Surplus carbon has no time to diffuse out, and is trapped in a supersaturated solution which takes up a body-centred-tetragonal structure. This phase is hard, brittle, highly stressed internally, and has a needle-like (acicular) appearance under the microscope.

The quenching rate required to suppress the austenite–pearlite transformation is known as the 'critical cooling velocity'. It is drastically reduced if alloying elements are present. Martensite formation starts at the so-called 'M_s' temperature, which is often about 350 °C.

Bainite is formed in certain alloy steels if the cooling rate is fast enough to suppress the austenite–pearlite change, but the material is not quenched below the M_s temperature. The properties of bainite are intermediate between martensite and pearlite.

Time—temperature—transformation curves (TTT curves)

From the foregoing discussion it is clear that the phases finally formed in a HAZ during cooling or subsequent heating will depend on time and temperature. TTT curves such as given in *Figure 2.21* show the time required for transformation to various phases at *constant temperature*, and, therefore, give a useful initial guide to

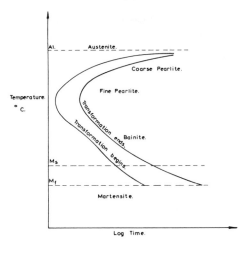

ISOTHERMAL   TRANSFORMATION   DIAGRAM.

Figure 2.21

likely transformations. A number of interesting points have emerged from such curves:

1 The time for austenite transformation is shortest at the 'nose' of the curve. If the cooling rate is sufficient to bypass this point, transformation takes place at a lower temperature. Hence a harder final phase is likely.
2 The final structure depends on the transformation temperature.
3 The shape of the curve, and its position on the time axis depends on the carbon and alloy contents. Carbon, manganese and nickel shift the curve bodily to the right without marked change of shape, whereas chromium and molybdenum alter the shape by shifting the austenite/pearlite transformation more to the right than the corresponding austenite/bainite transformation.

4 If the TTT curve is displaced to the right, the 'start-of-transformation' curve may not be reached even during relatively slow cooling, and as a result, martensite will be formed.

5 The temperature required for complete conversion to martensite (M_F), may be below ambient temperature (requiring refrigeration!).

Continuous cooling transformation diagrams (*CCT diagrams*)

The constant-temperature basis of TTT curves is obviously somewhat unrepresentative of welds, in which rapid continuous cooling takes place. More relevant information can be obtained from a CCT

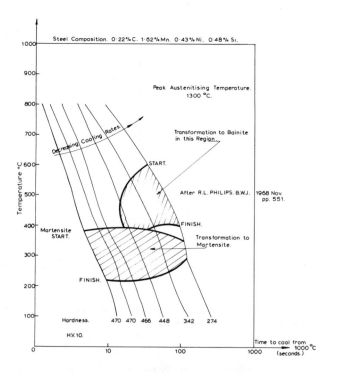

Figure 2.22

diagram, in which phase changes are tracked for a variety of cooling rates. To produce such diagrams for weld heating and cooling conditions clearly requires considerable experimental effort; in the example shown in *Figure 2.22*, special apparatus was used to simulate

HAZ cooling, and the transformation points were derived from both dimensional and temperature changes in the specimen. Few diagrams of this type are readily available.

Features of HAZ microstructures in steel welds

Grain growth

When a steel is held in the upper austenite range for a long time, the grains increase in size, by grain boundary migration and coalescence. This is unfortunate, as coarse-grained steels generally have poor

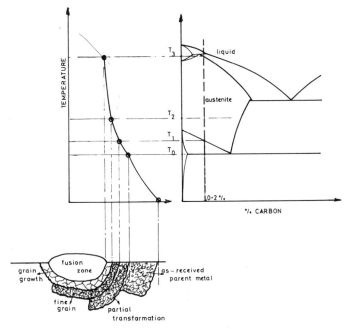

SCHEMATIC MICROSTRUCTURE OF 0·2°/. C   STEEL WELD

Figure 2.23

fracture toughness. The width of the coarsened zone depends on the welding process and parameters; for example in electroslag welding where the specific energy input is high (25–75 kJ/mm), wide HAZs are produced (often over 12 mm) and the grain structure tends to be coarse, both in the weld metal and in the HAZ. To improve notch toughness, therefore, 'normalising' procedures are often applied to reduce the grain size. In this treatment, the completed fabrication is heated uniformly to a point just above T_1 in *Figure 2.23*, and allowed to cool naturally in air. (The practical differences between normalising and stress-relieving should be noted. Not only is the required furnace temperature much higher (about 950 °C against 650 °C) but at the higher temperature the structure may not have sufficient strength to support its own weight.)

When the material is heated in the lower austenite range (T_1 to T_2), a fine-grained microstructure is produced. The upper temperature limit T_2 is controlled largely by grain-refining agents such as aluminium, niobium, vanadium and nitrogen. These form carbides and nitrides which prevent grain boundary migration but their usefulness reduces as the temperature is increased since they go into solution.

Precipitation, solution and overageing

High yield and creep resistance are obtained in some steels by uniform dispersion of precipitates throughout the matrix. Some of the strengthening may be lost if the precipitates are taken into solution in the high-temperature HAZ near the fusion line (just below T_3). In addition they may be 'overaged', at lower temperature, in which case the precipitates coarsen, and the strength is reduced. In such cases the original properties can only be restored by reheat treatment.

Partial transformation in the HAZ

In the remaining portion of the HAZ between T_0 and T_1, pearlite *in the original parent material* begins to transform to austenite. If subsequent cooling is fast enough, and the material is high in carbon and alloying elements, martensite may also be formed in this region.

Thus a composite picture can be constructed as in *Figure 2.23* of a microstructure which varies continuously from the unaffected parent material at T_0, through fine- and coarse-grained phases to the fusion line, and finally to the cast weld structure.

Tempered microstructures in multipass welds

In multipass welds, each weld bead is partially reheated and tempered by subsequent passes. This treatment greatly improves toughness, as carbon diffuses out of any martensite previously formed, leading to a tough microstructure consisting of ferrite and spheroidised carbides.

Here again it is clear that the welding engineer often has to make an unhappy choice between a procedure which gives excellent metallurgical properties (multipass), and an alternative procedure (single pass) which is economical but may give inferior properties.

2.6.2 Specifications for carbon and low-alloy steels

As some dilution will always occur in fusion welding, the first requirement for a weldable steel is that it should not contribute large quantities of troublesome constituents (elements or gases) to the weld pool. This limitation is obviously most severe for autogenous welds (100% dilution) in which case higher purities are specified.

The composition in other respects is dictated by behaviour in the HAZ. The aim is to reduce the risk of cracking by any of the mechanisms previously described. Some compromise will obviously have to be struck between an alloy composition which gives high strength (often valued by the designer) and simplicity and economy of fabrication (i.e. freedom from preheating restrictions, use of special welding processes, heat treatments, etc.). It is convenient to discuss the compositions recommended, element by element.

Carbon

The strength of steel increases with carbon content, often at the expense of toughness. Although some commercial steels have carbon contents of well over 0.5%, the risk of martensite formation, and the likelihood of solidification cracking at these levels inhibits their use in welded construction.

The maximum carbon levels permitted in the specification of various countries are in good agreement, namely:

Country	Carbon % max.	Comments
UK	0.3	BS 4360 C–Mn structural steels
	0.23	BS 1501 C–Mn pressure vessel steels
	0.15	BS 1501 $\frac{1}{2}$Cr–$\frac{1}{2}$Mo ...
USA	0.33	plain carbon & C–Mn structural steels
	0.26	in steels with 1% total of Cu, Ni, Cr, V, Nb
	0.25	in steels with 0.5% Ni, Cr or Mo
Sweden	0.25	in rimming steels
	0.22	in killed structural steels
Germany	0.26	in unalloyed pressure vessel steels

Manganese

This element is frequently used to maintain strength where the carbon level must be reduced for weldability. It also promotes martensite formation but to a lesser extent than carbon ($C_e =$ % C + % Mn/6 ... etc.). Hence a BS 4360 grade 55C carbon-manganese steel (yield strength 420 MN/m²) with 1.7% manganese maximum, can be butt welded without preheat up to 12 mm thickness provided that a low-hydrogen process is used.

Chromium, molybdenum and nickel

These elements also allow the strength to be maintained while the carbon level is reduced, although they also promote martensite formation. 9% Ni steel, for example, has a yield strength of 525 MN/m² with the carbon content held to 0.1%. (This steel also forms autotempered martensite in the HAZ, which is much less susceptible to hydrogen cracking than twinned martensite.) Nickel also promotes fracture toughness, and the 3½% Ni and 9% Ni grades are used for low-temperature service.

Sulphur and phosphorus

The damaging role of sulphur and phosphorus in solidification and liquation cracking has been described. The sulphur level in weldable structural steels is generally held below 0.05% and many investigators believe that it should be below 0.02% in steels which have a tensile strength higher than 700 MN/m². However, user demand for very low sulphur steel received something of a blow recently, when it was discovered that such steels with less than 0.01% S were unreasonably susceptible to hydrogen cracking. Present opinion[11] suggests that the manganese sulphide inclusions present in normal steels have a useful role hitherto unsuspected, in that they act as sinks for hydrogen. In a few cases these otherwise splendid steels had to be resulphurised before they could be used!

The limits on phosphorus are generally the same as for sulphur.

Some common specifications for weldable structural steel are given in *Table 2.5*. However reference 12 gives an exhaustive list of weldable specifications, which any designer would find invaluable.

Table 2.5

Designation	Type	Grade	C	Si	Mn	Ni	Cr	Mo		Yield strength (MN/m²)	Ultimate tensile strength (MN/m²)
Weldable structural steels (after BS 4360, Part 2, 1969)	Carbon steel	40A	0.22	—	—	—	—	—	—	—	400–480
		40B	0.20	—	1.50	—	—	—	—	210–230	400–480
		40C	0.18	—	1.50	—	—	—	—	210–230	400–480
		40D	0.16	0.1/0.5	1.50	—	—	—	0.10 Nb	225–260	400–480
		40E	0.16	—	1.50	—	—	—	—	225–260	400–480
	C–Mn steels	43AI	0.25	—	—	—	—	—	—	—	430–510
		43A	0.25	—	—	—	—	—	—	220–245	430–510
		43B	0.22	—	1.50	—	—	—	—	220–245	430–510
		43C	0.18	—	1.50	—	—	—	—	220–245	430–510
		43D	0.16	—	1.50	—	—	—	0.10 Nb	240–280	430–510
		43E	0.16	0.1/0.5	1.50	—	—	—	—	240–280	430–510
		50A	0.23	—	1.60	—	—	—	—	—	500–620
		50B	0.20	—/0.50	1.50	—	—	—	0.10 Nb	325–355	500–620
		50C	0.20	—/0.50	1.50	—	—	—	0.10 Nb	325–355	500–620
		50D	0.18	0.10/0.50	1.50	—	—	—	0.10 Nb	340–355	500–620
		55C	0.22	—/0.60	1.60	—	—	—	0.10 Nb	415–450	550–700
		55E	0.22	—/0.60	1.60	—	—	—	0.10 Nb	415–450	550–700

Table 2.5 (continued)

Designation	Type	Grade	C	Si	Mn	Ni	Cr	Mo		Yield strength (MN/m²)	Ultimate tensile strength (MN/m²)
Steels employed in low-temperature applications (after BS 1501, Part 2, 1970)	3¼% Ni steel (up to 1¼" thick)		up to 0.15	0.10/ 0.35	0.30/ 0.80	3.25/ 3.75	0/ 0.30	0.10	0.020 Al	260	450
	9% Ni steel		up to 0.10	0.10/ 0.30	0.30/ 0.80	8.75	0/ 0.30	0.20	0.020 Al	530	700
Creep-resisting ferritic steels (after BS 1501, Part 2, 1970)	1% Cr-½% Mo (up to 6" thick)	27	0.09/ 0.15	0.10/ 0.40	0.40/ 0.70	0.30	0.70/ 1.20	0.45/ 0.65	—	290	420–540
	1¼% Cr-½% Mo (up to 6" thick)		0.09/ 0.15	0.15/ 0.35	0.40/ 0.70	0.30	1.00/ 1.50	0.45/ 0.65	—	300–340	450–600
	2¼% Cr-1% Mo (up to 6" thick)	31	0.10/ 0.15	0.20/ 0.50	0.40/ 0.80	0.30	2.00/ 2.50	0.90/ 1.20	—	280	480–600
Quenched and tempered high-yield steels	Hy 80		0.18	0.15/ 0.35	0.10/ 0.40	2.00/ 3.50	1.00/ 1.80	0.20/ 0.60	0.03% V 0.02% Ti 0.25% Cu	550–700	—
	T 1		0.10/ 0.20	0.15/ 0.35	0.60/ 1.00	0.70/ 1.00	0.40/ 0.65	0.40/ 0.60	0.03–0.08% V 0.002–0.006% B 0.15–0.50% Cu	620–800	700–930
Quenched and tempered high-yield steels	Hy 100		0.20	0.15/ 0.35	0.10/ 0.40	2.25/ 3.50	1.00/ 1.80	0.20/ 0.60	0.03% V 0.02% Ti 0.25% Cu	690–830	—
	Hy 150		0.16/ 0.20	—	0.40/ 0.60	3.50/ 4.00	1.25/ 1.75	0.30/ 0.50	0.07–0.12% V	930–1070	1000–1140

(This table is adapted from BS 4360: 1972. Weldable structural steel, and is reproduced by permission of the British Standards Institution, 2 Park Street, London W 1A 2BS)

2.6.3 Stainless steels

These steels are valued for corrosion resistance, which is due to a chromium oxide surface film, and for their high-temperature strength. A variety of metallurgical structures can be obtained depending on chemical composition. These are conveniently illustrated in terms of the Schaeffler diagram (*Figure 2.24*) in which elements are assessed according to their ability to promote austenite phase (C, Ni, Mn) or ferrite phase (Cr, Mo, Nb, Si) in the weld metal. The version of the diagram illustrated also indicates the character of welding problem which can arise in each part of the composition range.

Austenitic stainless steel is probably the most popular grade, and is based on a composition which is austenitic at ambient temperature (roughly 18% Cr, 8% Ni). It is usual, however, to adjust the weld metal composition so that between 4% and 10% ferrite is formed, as fully austenitic weld metal is susceptible to solidification cracking due to impurity films. Ferrite acts as a useful sink for harmful impurities, but more than 10% lowers corrosion resistance. It also leads to the formation of a hard brittle phase ('sigma') if the metal is reheated in the range 500 to 900 °C in service. As shown in *Figure 2.24* therefore, the range of weldable compositions is rather narrow. (N.B. In many material specifications such as the EN 58 series, the permitted composition limits are wide enough to be outwith the safe range if 100% dilution occurs. Closer specification of composition or suitable filler metal additions are therefore essential.) The Schaeffler diagram is also useful for the assessment of dissimilar metal joints. In many piping systems, for example, different materials are used in different parts of the system according to the fluid temperature existing at that part. The designer may, therefore, call for a joint between, for example, a $2\frac{1}{4}$% Cr–1 Mo tube (A), and an austenitic stainless steel tube (B). The separate compositions A and B are first located on the diagram, *Figure 2.25*, as shown. If an autogenous tube butt welding process is used (even for the root run) the resulting composition will be roughly at the point midway between the two (D) and will be liable to crack. If a filler metal or MMA electrode is used which has a composition (allowing for losses) corresponding to C, and the dilution is arranged to be between 42% and 62% (see *Table 2.1*) a satisfactory joint can be produced corresponding to composition 'E'.

For severely corrosive environments where the ferrite level must be reduced to zero, it is clear that high-purity materials must be specified. In these 'fully austenitic' deposits it has been argued that solidification cracking can be attributed to silicon, and that the

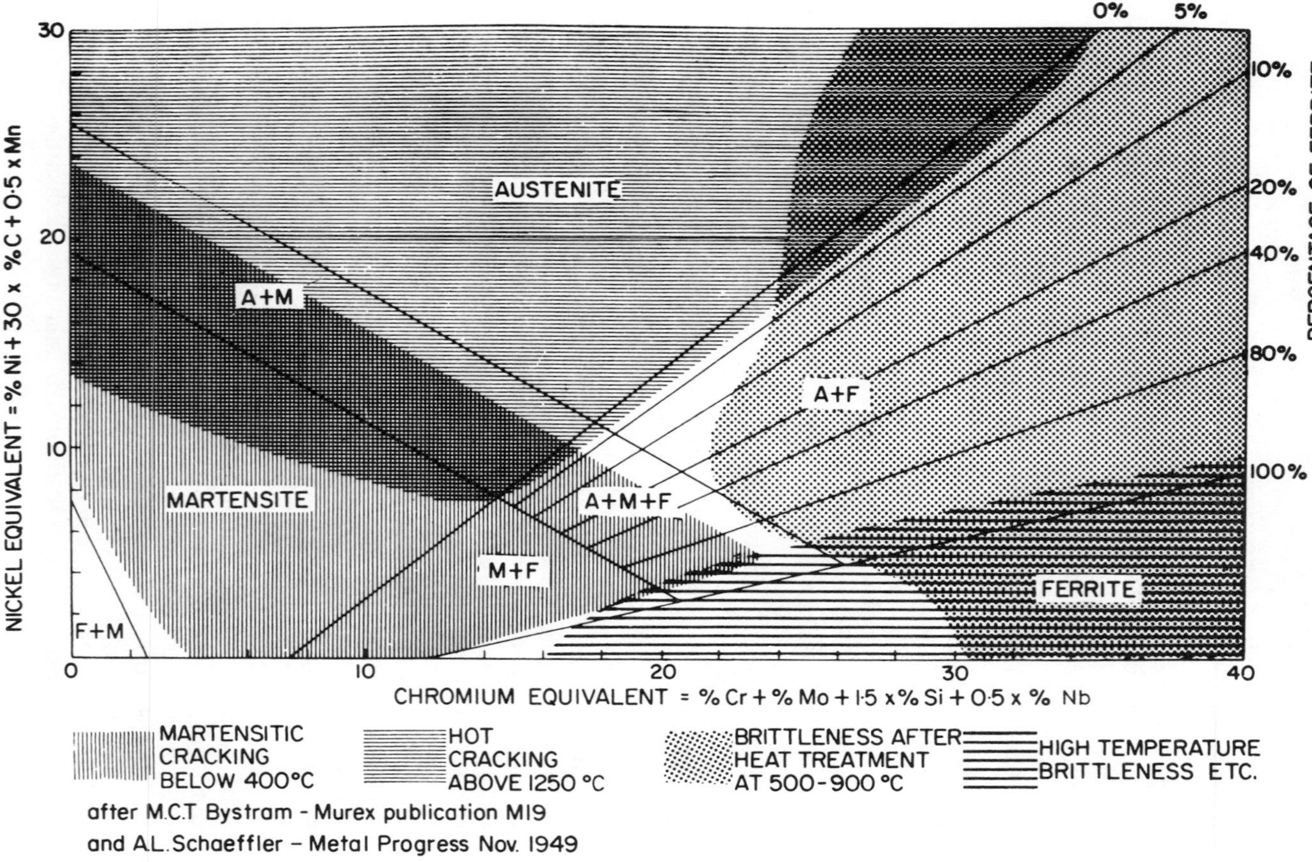
0%
5%
10%
20%
40%
80%
100%
PERCENTAGE OF FERRITE
30
20
10
0
NICKEL EQUIVALENT = % Ni + 30 x %C + 0·5 x Mn
AUSTENITE
A+M
A+F
MARTENSITE
A+M+F
M+F
F+M
FERRITE
10
20
30
40
CHROMIUM EQUIVALENT = % Cr + % Mo + 1·5 x % Si + 0·5 x % Nb
MARTENSITIC CRACKING BELOW 400°C
HOT CRACKING ABOVE 1250 °C
BRITTLENESS AFTER HEAT TREATMENT AT 500-900 °C
HIGH TEMPERATURE BRITTLENESS ETC.
after M.C.T Bystram - Murex publication M19
and A.L. Schaeffler – Metal Progress Nov. 1949

Figure 2.24 *Schaeffler diagram*

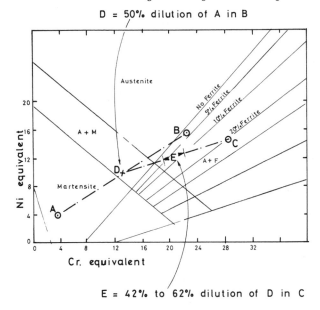

Figure 2.25

ratio $C/\sqrt{\text{Si}}$ should be maintained above 0.22 in order to avoid cracking[13]

Weld decay and stabilised stainless steels

If stainless steels are held in the 400–800 °C range for a time (either during welding or in service), chromium carbide may be formed and precipitated at grain boundaries. These areas will, therefore, be depleted of chromium and will become open to corrosive attack (weld decay). A number of methods can be used to reduce the problem:

1. Restrict the carbon content to 0.03% (minimising carbide precipitation). This approach is used in AISI type 304L, and is probably the soundest method.
2. 'Stabilise' the carbon by the addition of niobium and titanium which form carbides more easily than chromium. (N.B. Titanium transfers poorly across the arc, and is not used for electrode fillers.) This approach is used for type 321 (EN 58 B & C) and type 347 (EN 58 G).

3 Add molybdenum (usually with Ti and Nb), as in type 316 and 317 (EN 58 H and J). This reduces the precipitation tendency.
4 'Solution treat' at 1050 °C after welding to take the carbide phase back into solution.

Martensitic and *ferritic* stainless steels are somewhat cheaper (nickel being an expensive element), but are more difficult to weld. The martensitic grade is hard and brittle and extremely susceptible to hydrogen-induced cracking. In the ferritic grades the toughness of the HAZ is poor, due to grain growth during welding. These steels are, in fact, superior to austenitic stainless steels for certain applications; they are less susceptible to transgranular stress corrosion cracking by chlorides, and are resistant to high-temperature oxidation and attack by sulphur-bearing compounds. Typical welding procedures are given in *Table 2.6*.

Table 2.6   WELDING OF STAINLESS STEELS

Steel	Typical composition	Welding problem	Practical solution
Austenitic stainless steels	18% Cr–8% Ni 25% Cr–20% Ni 25% Cr–12% Ni 15% Cr–35% Ni	Solidification cracking in weld metal	Maintain 4–10% ferrite phase in weld metal. Keep $C/\sqrt{Si} > 0.22$ in fully-austenitic weld metals
		Weld decay	See text
	18% Cr–13% Ni–1% Nb	Reheat (stress relief) cracking in the HAZ	Rapid heating through the critical temperature range. Use of electrodes having lower hot strength. Grind welds to remove stress concentration effects.
Martensitic stainless steels	12–16% Cr–0.3% C (max)	Hydrogen-induced cracking in the HAZ	Preheat 200–400 °C and post heat at 750°C low-hydrogen electrodes OR austenitic stainless steel electrodes. See reference on *Figure 2.24*
Ferritic stainless steels	16–30% Cr–0.1% C	Low toughness in grain-coarsened HAZ	Preheat 200 °C. See reference on *Figure 2.24*

2.6.4 Aluminium and its alloys (see Table 2.7)

Some of the newer fusion processes (e.g. GTA, plasma-arc, GMA and electron beam) are particularly suited to the welding of aluminium alloys. These processes operate at a high energy density, use no flux and therefore relate better to the thermal properties of aluminium, to its affinity for oxygen and resulting tough oxide film (see page 21) and to the problems of gas entrapment in the rapidly freezing weld. Porosity is a major problem, and one cannot emphasise too strongly the need for clean practice and uncontaminated shielding gas. Apart from porosity, two other problems frequently arise, namely, solidification cracking and HAZ softening.

Cracking

Al–Mg–Si and Al–Mg–Zn alloys are susceptible to the cracking mechanism described on page 89 for Al–Mg alloys. All these alloys are therefore welded using nonmatching fillers (namely, Al–5% Mg for Al–Mg and Al–Mg–Zn alloys, and Al–10%Si for Al–Mg–Si alloys). In the case of Al–Mg–Zn, the recommended filler material will have a lower strength than the parent material, and considerable efforts are being made to find a weld metal which can match the parent metal, perhaps in conjunction with a post-weld ageing treatment.

Softening

The non-heat-treatable alloys (including commercially-pure aluminium, Al–2–5% Mg, and Al–1¼% Mn) are often used in a work-hardened condition, as otherwise they have a rather low yield strength. In such cases, annealing will occur in the HAZ, leading to a soft zone.

Most of the heat-treatable alloys (including Al–Mg–Si and Al–Mg–Zn) derive their strength from precipitation hardening of the matrix, brought about by carefully controlled heat treatment. As in the case described on page 108, the precipitates will be taken into solution by welding, producing a softened HAZ. The original strength can be restored by high-temperature solution treatment and re-ageing, which is not always practicable for completed structures. However, there are some alloys (notably Al–4% Zn–1% Mg) which age-harden 'naturally', that is at ambient temperature over a period of days or weeks. Hence the HAZ properties will be regained if

Table 2.7 TYPICAL WELDING PROBLEMS AND SOLUTIONS IN ALUMINIUM ALLOYS

Alloy type	Typical composition	Typical welding problems	Usual solution
All types		Porosity formation	Clean welding procedures, and pure shielding gas
Non-heat-treatable			
Commercially-pure aluminium 99.5% Al		Fe, Si impurities can form intergranular films in weld metal & HAZ, lowering corrosion resistance	Use high-purity filler metal, (99.8% Al)
Al–Mn	Al–$1\frac{1}{4}$% Mn	Porosity (see above)	
Al–Mg	Al–2/5.5% Mg	Solidification cracking in weld metal	Use Al–5% Mg filler wire
Heat-treatable			
Al–Mg–Si	Al–$\frac{3}{4}$% Mg–1% Si	Autogenous welds exhibit solidification cracking	Use Al–10% Si filler wire
	Al–6% Zn–2.8% Mg	Autogenous welds exhibit solidification cracking	Use Al–5% Mg filler wire
Al–Zn–Mg	Al–2.75/3.75% Zn 1.5/2.5% Mg	Autogenous welds are uncracked	Use Al–2% Zn–4% Mg, OR, Al–5% Mg filler wire

sufficient time is allowed. Of course if Al–5% Mg filler wire is used in order to avoid the cracking problems previously mentioned, the *fused zone* will have a lower yield strength than the base metal. In both types of alloys, therefore, it is important to minimise the width of the HAZ and the time at high temperature. From Section 1.2, it is clear that a low specific energy welding process achieves both of these aims, and incidentally minimises distortion which is a considerable problem in welded aluminium structures.

REFERENCES

1. ROLLASON, E. C., *Metallurgy for Engineers*, Edward Arnold & Co., London
2. JONES, J. E., *et al.*, 'The application of recent welding techniques to heavy fabrications', *J. West of Scotland Iron and Steel Inst.*, **74** (1966)
3. DEARDEN and O'NEILL, 'A guide to the selection and welding of low-alloy structural steels', *Trans. Inst. Welding (U.K.)*, **3**, No. 10, 203–214 (1940)
4. ITO, Y. and BESSYO, K., I.I.W. Document IX/576/68, 6–8
5. ITO, Y. and BESSYO, K., I.I.W. Document IX/631/69 (July)
6. BORLAND, J. C., 'Cracking tests for assessing weldability', *British Welding J.*, **7**, 630 (1960)
7. BAILEY, N., 'Welding carbon manganese steels', *Metal Construction*, **2**, No. 10, 442 (1970)
8. PRATT, J. L., 'The automatic welding of BS 968 (1962) steel', *British Welding J.*, **13**, No. 9, 513 (Sept. 1966)
9. JUBB, J. E. M., CARRICK, L., and HAMMOND, J., 'Some variables in lamellar tearing', *Metal Construction*, **1**, No. 25, 58–63 (Feb. 1968)
10. NAKAMURA, H., NAIKI, T., and OKABAYASHI, H., 'Fracture in the process of stress relaxation under constant strain', *Proc. 1st Int. Conf. on Fracture*, Vol. 2, 863–878 (1965)
11. SMITH, N. and BAGNELL, B. I., 'The influence of sulphur on the HAZ cracking of C–Mn steel welds', *Metal Construction*, **1**, No. 25, 17–24 (Feb. 1968)
12. STOUT, R. D., and DOTY, W. D. O., *Weldability of Steels*, Welding Research Council, New York (1971)
13. POLGARY, S., 'The influence of Si content on cracking in austenitic stainless steel weld metal', *Metal Construction*, **1**, 93–129 (Feb. 1968)
14. ROBERTS, D. K., and WELLS, A. A., 'A mathematical examination of the effect of bounding places on the temperature distribution due to welding', *British Welding J.*, **1** (Dec. 1954)
15. DOLBY, R. E., 'Weldability of low-carbon structural steels', *Low-carbon Structural Steels for the Eighties*, Inst. of Metallurgists Conf., Spring 1977, IIIB, 1–11 (1979)

BIBLIOGRAPHY

Lamellar tearing in welded steel fabrication, The Welding Institute (1971).
LINNERT, G. E., *Welding metallurgy* Textbooks, Vols. 1 and 2, American Welding Soc., New York (1965)

PART II. DESIGN FOR STRENGTH

Next to 'heat-proof', I suppose that 'unbreakable' is one of the most useful words in advertising.

J. E. GORDON

Chapter 3

Identifying the problem

3.1 The features of welded construction that influence service performance

It is understandable that the failures which have occurred in welded construction are better known than the successes. However, not all these failures have been directly attributable to welds as such, nor have they always been caused by 'bad' welding, in the sense of poor workmanship. Nevertheless, the memory of certain unexpected and dramatic collapses seems sometimes to have inspired an unfortunate

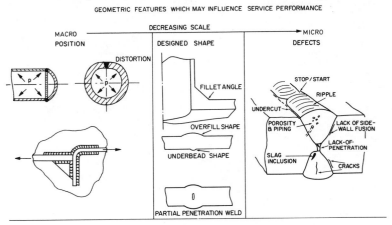

GEOMETRIC FEATURES WHICH MAY INFLUENCE SERVICE PERFORMANCE

Figure 3.1   Geometric features which may influence service performance

and somewhat hysterical impression that welding causes strange and unpredictable behaviour in structures, and that the reasons cannot be understood. Such an attitude is a hindrance to good engineering design and this chapter will have succeeded if it encourages the designer in a more positive philosophy. Let us accept, therefore, that any unwelcome behaviour in welded components should be attributable to identifiable effects, such as geometrical features (see *Figure*

123

3.1), changes in the physical properties of the materials or residual stress fields. These features form headings for the following discussion.

3.1.1 Geometry

The position of the weld in the structure may be enough in itself to ensure that failure occurs in or near the joint. Through force of circumstance, welds are often placed at those points in the structure which already experience maximum stress—extreme fibres in bending, or load transfer points. Whether the weld quality is good or bad,

Figure 3.2   Crack between closely spaced fillet welds (holes drilled at crack ends). (Courtesy: Babcock & Wilcox (Operations) Ltd)

these are, therefore, the most likely positions for the initiation of failure. In other cases, trouble is caused by congested arrangements of intersecting plates and gussets which form stress concentrations. Closely spaced welds not only increase the risk of failure in service, but can stimulate cracking during fabrication as shown in *Figure 3.2*. It is particularly distressing to come upon failures which have been caused by attachments of a temporary or otherwise trivial nature—

lifting lugs, jacking pads or the like. In *Figure 3.3* a pressure-vessel fatigue failure is shown, stemming from a man-door hinge which has been attached at the point of maximum stress on the torispherical head. Distortion may also help to raise the stresses in the joint area. For example, a perfectly rolled cylinder will usually be flattened or

Figure 3.3   Pressure-vessel fatigue failure

peaked at the closing weld due to angular distortion. Under internal pressure loading, additional bending stresses will be developed at the seam. In thin, compressively loaded structures, dents and undulations in the weld area caused by distortion will stimulate additional buckling stress.

Leaving aside the position of the weld, in the structure its shape may have an influence on failure, particularly in creep or under fatigue loading. Considerable variations in the shape of fillet and butt welds are usual, but even geometrically regular shapes will create a large concentration of stress.

On a finer scale, the designer needs to be aware of the variety of irregularities and defects which may arise, ranging from surface ripple to large cracks as shown in *Figure 3.1*. Such defects can only be eliminated at great cost, and in any case they are not invariably detrimental to service performance. In general sharp crack-like defects should give more cause for concern than rounded flaws such as porosity. Certain cracks which develop during fabrication may not be considered to endanger the structure (see Chapter 5). They should nevertheless be investigated carefully, because weak metallurgical structure is often indicated.

3.1.2 Variation in properties

It follows from the discussion in Chapter 2 that the physical properties of the joint material will vary considerably from the central weld metal to the unaffected parent material. Unless the chemical composition is substantially different, the elastic modulus should be unaffected, but other properties such as yield strength, ductility, fatigue and creep strength may vary enormously. This can lead to a 'metallurgical notch' in which stress or strain is concentrated in stronger or weaker parts of the weld zone, even in the absence of a geometrical notch. Mostly, the mechanical properties are assessed in terms of standard tensile, bend or notch impact tests, which are on the whole insensitive to local alterations in property. This would not matter if the service behaviour of real structures was similarly insensitive but, especially with respect to creep behaviour at elevated temperatures, or crack propagation resistance, the standard tests offer little guidance. The indentation hardness test is sometimes useful for the identification of troublesome zones.

Trouble may occur in fluctuating-temperature service when adjacent materials have greatly different thermal expansion coefficients. As an example, thick austenitic steel weld cladding may be torn off a ferritic steel backing especially if the interface between the two is brittle.

Corrosion thrives on adjacent microstructural differences, even where the chemical match is good[1]. Hence weld joint areas may require special corrosion protection. A vertical storage tank collapse is known to the authors in which a butt weld had apparently been gouged away at the weld toes from the adjacent plate by the action of distillery waste products.

As a general rule, grain refinement improves the properties in most respects, whether it is brought about by multirun welding, or post-weld normalising heat treatment.

3.1.3 Residual stress

In most fusion welding procedures, tensile residual stresses of yield point magnitude are generated in the welding direction (see Section 1.4). These stresses can be partially or wholly removed mechanically or thermally (see Section 4.3) and there is no doubt that stress-relieved structures have a much better record than non-stress-relieved ones. Obviously any stress-operated failure mechanism will put the unrelieved structure at risk. In particular, one might mention brittle fracture, fatigue crack propagation, and stress corrosion which have led to failure in association with residual stress.

It is a common experience that components which have cracked or worn, and have been repaired or built up by welding, do not survive the service loading for long, even when the parts have been carefully remachined. Residual stress is very often the culprit, and reference 2 includes a long and depressing catalogue of such mishaps.

3.1.4 Illustrations

The collapse of the King's Bridge in Melbourne[3] followed a classic pattern which has been repeated elsewhere. Hydrogen-induced cracking during fabrication provided initiating defects which were subsequently extended by fatigue loading, leading to brittle fracture of a main supporting girder. The cover plate detail which caused the trouble is shown in *Figure 3.4*.

In terms of the geometrical features discussed in *Figure 3.1*, the weld at the fracture origin was positioned on a tensile extreme fibre, the fillet profile being transverse to the service loading, and to the residual stress system generated by the longitudinal cover plate welds. In addition, the related effects of carbon equivalent, preheat and low-hydrogen practice had not been taken seriously by the fabricator, and as a result, cracks of appreciable size were formed at the toe or the root of the transverse welds.

It is clear, therefore, that the HAZ toughness of the joints was poor, but even the unaffected parent plate material had an inadequate fracture toughness in terms of minimum impact energy absorption requirements. Hence the final brittle fracture only required a moderately heavy load, provided by a large road trailer, and a low ambient temperature (only about 4 °C).

In contrast to the previous bridge failure, the metallurgical properties of the fabricated lorry axle shown in *Figure 3.5* were not to blame for the eventual failure, as the axle had apparently been furnace-normalised after welding (see page 107).

However, the basic joint design was poor in the light of an

obvious fatigue loading in service, as an incomplete penetration butt weld had been specified. The unwelded portion was in the end even greater than designed, as there was poor side-wall fusion in the root area where the V-groove edge preparation had been too narrow for

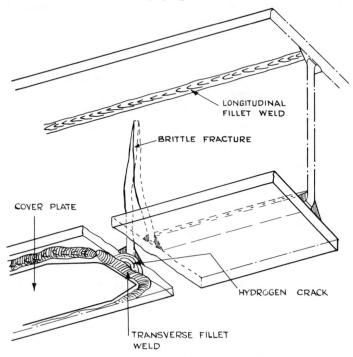

Figure 3.4   King's Bridge girder failure

the heavy deposition welding process (see page 44). The weld overfill was ground flush, presumably for appearance's sake, but even if it had been left as-welded, any stress concentration effect in this area would have been surpassed by the central defect.

The tensile principal stresses at the point of fatigue crack initiation were estimated to be up to $90\,MN/m^2$ and the crack spread incrementally for a time (concentric ring pattern), until the final unstable fracture occurred (radial lines).

3.2 The role of mechanics in the design process

It is true that many engineering failures are caused by details which could have been readily condemned at the design stage, without so much as a line of calculation. It is important, therefore, that the welding designer should develop the ability to spot such details, perhaps by studying references like reference 2 in conjunction with *Figure 3.1.*

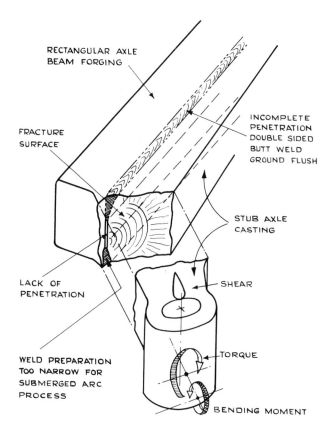

RECTANGULAR AXLE BEAM FORGING

INCOMPLETE PENETRATION DOUBLE SIDED BUTT WELD GROUND FLUSH

FRACTURE SURFACE

STUB AXLE CASTING

LACK OF PENETRATION

SHEAR

WELD PREPARATION TOO NARROW FOR SUBMERGED ARC PROCESS

TORQUE

BENDING MOMENT

Figure 3.5   Lorry axle fracture

However, in most cases the engineer will wish to make some numerical estimate of the likelihood of failure, either by relating the present design to previous designs, or by comparison with the material strength properties. This comparison is best made through the descriptive or characterising parameters provided by applied mechanics (e.g. stress, strain energy, energy rate, stress intensity, etc.). The structure of such an approach is illustrated in *Figure 3.6.*

The general term 'failure' is too imprecise for engineering purposes, as failure can occur in practice by a variety of different processes, which require very different mechanics' descriptions. For example, failure may occur:

1 By excessive elastic, permanent, or creep deformation.
2 As a result of unstable load/deformation behaviour (due to

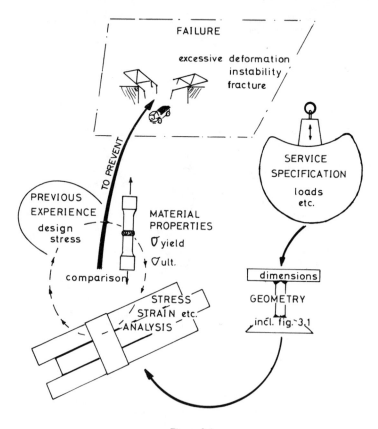

Figure 3.6

unstable plasticity in the material, or geometric or energy instability in the structure, as in buckling).

3 By fracture (separation of the component into parts, whether by an incremental process such as fatigue, or by a monotonic process which may or may not be stable such as 'brittle' fracture, creep rupture, stress corrosion cracking etc.).

Apart from the question of distortion and buckling, the calculation of deformations is not especially difficult in welded structures, nor is the 'excessive deformation' failure mode particularly troublesome. Fracture, on the other hand, is a significant failure mode for such structures, and the recent development of an applied mechanics understanding of fracture (or more precisely crack propagation) is therefore welcome. For these reasons Chapters 4, 5 and 6 concentrate on design against fracture, with little attention paid to deformation.

In trying to assess the risk of failure, the starting point is provided by the *service specification* which includes the magnitude, rate of application, and frequency of occurrence of loads on the component, features of the environment surrounding the structure (it may be corrosive, very cold or very hot), and the importance of the structure (would failure be catastrophic to life or property, or merely inconvenient?). Of course in many structures the loads are difficult to estimate with any certainty, for example, in vehicles, or excavation machinery, and one must rely very much on information from previous designs.

The specified loads and structural dimensions then form the input to an applied mechanics calculation of stress, strain or some other suitable characterising parameter. The *geometry* (which should include the features in *Figure 3.1*) may be variable in the design context. Component shapes which are designed for simple manufacture, or to fulfil a function other than the support of loads, often turn out to be extremely difficult to analyse (for example rectangular box tanks), and in many cases the shape may have to be altered to conform to a geometry which can be analysed without too much trouble.

In the final step shown in *Figure 3.6*, the value of *stress* calculated for the designed component is then compared with data derived from *previous experience*, or with a *strength property*, established by testing what one hopes is a representative sample of the material to be used. This approach certainly works well for yielding (comparison of applied stress with yield strength) but can be misleading for fracture, simply because the mechanism of fracture in a simple tensile test is usually different from that occurring in real structures.

REFERENCES

1. STEWART, D., and TULLOCH, D. S., *Principles of Corrosion and Protection*, Macmillan (1968)
2. *Fatigue fractures in welded constructions*, International Institute of Welding
3. *Failure of the King's Bridge, Melbourne*, Official report of the Royal Commission, Victoria (1963)

BIBLIOGRAPHY

PFLUGER and LEWIS, *Weld imperfections*, Addison Wesley (1968)
'The significance of defects in welds', *Proc. 2nd Conf.*, The Welding Institute (1968)
THIELSCH, H., *Defects and Failures in pressure vessels and piping*, Reinhold (1965)

Chapter 4

Stress analysis

4.1 Basic stress analysis

The term 'stress analysis' is frequently used, not only to cover the analysis of stress but the complete behaviour of a body including, on occasions, the distribution of load, stress displacement, strain and temperature throughout the body when some external actions such as load, displacement or temperature act upon the body. The rather obvious justification for doing stress analysis is that it is one of the main activities that can give a designer a quantitative yardstick as to the efficiency of a load-carrying component and the possibility of its failure in service. It is thus a necessary stage in the rational design of most engineering components but is usually a fairly approximate tool unless the loads and the geometry of the component in question are simple and well defined. This last point must be kept firmly in mind and 'stress analysis' needs frequently to be liberally seasoned with engineering judgement.

4.1.1 Fundamental concepts

For the unwary the field of stress analysis is confused by much of the terminology and techniques in such a way that fundamental concepts are sometimes lost. There are, in fact, only three basic principles.

1 EQUILIBRIUM   When a body is in a state of rest (or uniform motion) the loading both internally and externally must be in equilibrium. The application of this principle is described as 'statics' analysis.
2 COMPATIBILITY   When a body deforms under load it must do so in such a way that movements of adjacent elements of material are compatible with each other and with any support conditions, i.e. they must fit together in their deformed state without gaps or overlapping. This principle therefore reduces to

the application of geometrical rules and is often described as 'kinematics' analysis.

3 STRESS/STRAIN RELATIONS   The loads and deformations must be related in such a way that they satisfy the physical relationships between stress, strain, time and temperature for the material concerned. These are sometimes termed the Constitutive relations since they embrace more than stress and strain.

If, somehow, mathematics could be made simple enough for all to understand, then theoretical stress analysis would in a sense be quite trivial, since it is based on so few basic premises. When the above three conditions can be satisfied throughout a body, we have what is usually termed, an 'exact' solution. However, the problem as idealised for analysis, i.e. the mathematical model, is not always identical with the real physical problem and one should be wary of speaking too glibly about exact solutions. It follows that when any of the above conditions are not satisfied (or not completely satisfied) we have an approximate solution. Most of stress analysis is concerned with approximate solutions.

There are, however, alternative but similar routes to a solution. For example energy methods can be used in conjunction with the three basic requirements in what turns out to be simply an integrated form of the equilibrium and compatibility conditions linked by the relevant constitutive relation, the integration being throughout the volume of the body. In addition a number of energy principles have been established mainly for elastic materials which when developed for the appropriate constitutive relationship allow solutions to be found by optimisation of the energy, using either equilibrium *or* compatibility. Usually these are employed to arrive at approximate solutions but they have the advantage over some other approximate methods in that it is sometimes possible to show that the answer obtained (for a load or a displacement) is a 'bound'. That is to say it is known to be higher or lower than the exact answer. Two calculations then may allow the exact answer to be bounded, hopefully fairly closely. It should be noted that bounding of local stresses, as distinct from generalised quantities such as load or characteristic displacement, is not easy. It is also worth noting that in the use of the energy principles three conditions are still required. At its best then an energy approach may become identical to an exact solution if you try hard enough and obey all the rules, but in practice it is usually used as a powerful approximate tool.

It is unfortunate in a way that with the passage of time, short cuts to solutions (particularly of simple problems) have led to techniques

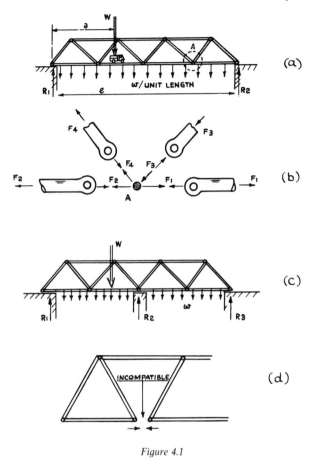

Figure 4.1

where the basic principles are difficult to distinguish but they are nevertheless present. The problem of the bridge in *Figure 4.1a* makes a convenient illustration of the principles discussed above.

Equilibrium

It seems obvious enough that the important components of the reactions from the supporting piers will be vertical as shown. Of course horizontal movement (due to temperature change for example) and friction could cause horizontal forces at the piers.

However, in practice the supports will usually be arranged to avoid such forces—one end may be located on a roller-type support free to move horizontally. The principle of EQUILIBRIUM states that unless the *vector* sum of the forces acting on a particle is zero, the particle moves. If the bridge is treated as a whole, then the vertical force W on the bridge, plus its selfweight must be equal to the sum of the vertical reactions. Or,

$$W + wl - R_1 - R_2 = 0$$

Furthermore, the sum of the *moments* of the forces acting on the particle must be zero (if this were not so the bridge would spin in mid-air). Taking moments about the left-hand end gives

$$Wa + wl\frac{l}{2} - R_2 l = 0$$

Solution of these equations for the two unknowns R_1 and R_2 is straightforward and gives

$$R_2 = \frac{Wa}{l} + \frac{wl}{2}$$

$$R_1 = W\left(1 - \frac{a}{l}\right) + w\frac{l}{2}$$

Having established all of the external forces on the bridge, one can now examine the forces on each component. The equilibrium principle may be applied to a 'particle' such as the pin joint at A in *Figure 4.1b* and then to each member until the forces *within* the structure are determined. In practice one would start with the simplest joint such as one at a pier in this case. Notice that pin-ended members like those shown here, when loaded at the pins without any transverse actions along their length, can only support a load acting along the axis of the member. Notice too that to analyse the framework in this way we have neglected the distributed (selfweight) load shown on the bottom horizontal members in *Figure 4.1a*. However, that layout was only for convenience in calculating the reactions. The problem can be easily avoided by assuming the weights of each member to be concentrated (and applied) at the ends, i.e. the pins.

When all the forces can be determined from equilibrium alone as in this case, the problem is said to be 'statically determinate' or 'isostatic'. To obtain a *complete* solution (strain and displacements) it is of course necessary to employ the other two principles. The simplicity of statically determinate analyses provides an incentive to design bridges and other components in terms of a pin-jointed structural model. If, in practice, the joints are not pinned but welded, analytical problems may arise unless we use some tricks of detail design as shown on page 188 to make the joint behave like a pin.

Many of the equilibrium-based techniques for force analysis (polygon of forces, 'methods of sections', etc.) are described in references 1–3.

Determinateness

Consider *Figure 4.1c* where the bridge is supported on *three* piers instead of two. Applying the earlier arguments:

$$W + wl - R_1 - R_2 - R_3 = 0$$

$$Wa + \frac{wl^2}{2} - R_2 \frac{l}{2} - R_3 l = 0$$

Now we have *three* unknowns and only two equations. All the reactions cannot therefore be determined. The reason for this anomalous behaviour lies in the false assumption that the 'particle' or bridge is rigid. If this were so, then marginal differences in the pier heights would mean that the bridge would be supported entirely by two of the three piers. All three reactions cannot then be determined until the deformation of the system is also considered. The principle of equilibrium is insufficient. This state of affairs arises frequently in analyses and is described by the term, 'statically indeterminate' or 'hyperstatic'. The test for indeterminacy is that the deformation of the component influences the loads on it.

Generally having analysed for the forces on and within the structure we are not finished with equilibrium. We must use it to establish relationships between the stresses on different planes within the material under load. This is considered later under the heading 'Stress Transformations' on page 161.

Compatibility

Referring to *Figure 4.1a*, the bridge (as a whole) is connected to the supporting piers. Therefore if the piers sink under load, so does the bridge. What is more to the point, in *Figure 4.1c*, compatibility tells us that three points on the bridge, i.e. at the supports, are on the same level or related levels, and this fact can be and must be used to establish the deformed shape of the bridge.

Similarly, turning from the whole bridge to components within the bridge, in general if the tie bars in *Figure 4.1d* are to remain connected, restrictions must be placed on their changes in length under load.

An analogous situation exists with deformations as was mentioned with forces. In some simple structures, if the deflection is known at one point then the displacement (or strain) can be found everywhere

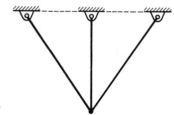

Figure 4.2

using compatibility alone. Such a structure is said to be *kinematically determinate*, for example the simple three-bar pin-ended structure shown in *Figure 4.2* is statically indeterminate if a load *P* is applied in any direction at the common point but is kinematically determinate if the displacement is known at that point, i.e. the changes in length and strains can be found for each member from geometrical rules (compatibility) alone.

Compatibility also places restrictions on the deformations of material *within* a component, and this will be considered under the heading 'Strain Transformations' on page 166.

Constitutive relations

The final step in analysing our bridge, is to link the loads in the members to the deformations produced. To do this we must know the physical relationships between stress and strain for the bridge material. The relations may be elastic, anelastic, linear, nonlinear, time dependent, isotropic, etc. In addition, strain and temperature in a component are connected via the coefficient of thermal expansion. A few typical stress/strain relationships are shown in *Figure 4.3*, including both practical and idealised behaviour. The large bulk of information available from stress analysis refers to linear elastic material. Because of mathematical complexity it is only in recent years that relations other than linear elastic have been considered but it is now possible to obtain answers for a wide range of problems for different material idealisations. Despite the fact that actual material behaviour departs significantly from the common linear elastic behaviour it is rather fortuitously the case that linear elastic solutions are still useful and can be used meaningfully in design of components for limited plasticity, fatigue conditions and even creep or fracture. Indeed the most powerful present-day

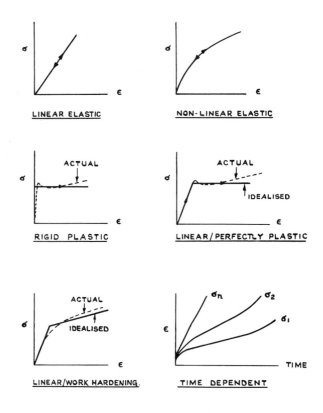

Figure 4.3   Constitutive relationships

approaches to these problems invariably utilise the linear elastic solution in some shape or form.

Mathematical equations can be fitted to any of the relationships shown in *Figure 4.3*. The simplest and commonest is the linear elastic case where for uniaxial loading $\sigma = E\varepsilon$. The constant E is known as Young's modulus. For interest, a number of auto-graphically recorded stress–strain curves for a variety of materials are given in *Figure 4.4*.

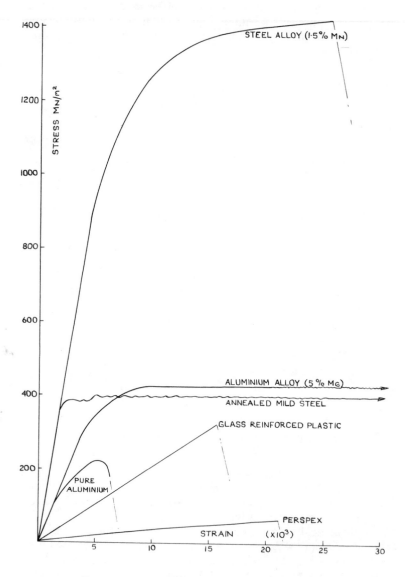

Figure 4.4   Stress–strain curves for various materials

4.1.2 Solutions for simple loading systems

The following simple formulae which are used to estimate nominal stress levels in welds and components are well known, but it is easy to misapply them unless one is clear about their origins. First some definitions.

Normal stress σ, load acting normal to a plane per unit area

Shear stress τ, load acting parallel to a plane per unit area (friction forces produce shearing action)

Normal strain ε, change in linear dimension

Shear strain γ, change in angle or shape.

Direct action

The normal or direct stress on any plane at right angles to the load is uniform and has the value

$$\sigma = \frac{P}{A} \tag{4.1}$$

which from equilibrium will be the stress on any cross section along the length of the bar (*Figure 4.5*).

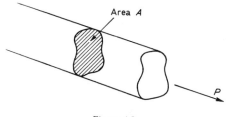

Figure 4.5

For linear elastic material the stress–strain relationship is usually denoted by Young's modulus $E = \sigma/\varepsilon$. Then the strain is given by

$$\varepsilon = \frac{P}{A}\frac{1}{E} \tag{4.2}$$

Equation (4.1) is derived from equilibrium and equation (4.2) from the stress–strain relationship. To obtain the elongation of a length

L of the shaft, a compatibility relationship is required; in this case the trivial one:

$$\text{elongation} = \int_0^L \varepsilon \, dx = \frac{PL}{AE}$$

Bending action

Bending loading situations can also be easily used to illustrate the basic principles of stress analysis. Consider a beam whose cross-sectional dimensions are small compared with its length under the action of a constant bending moment M as shown in *Figure 4.6*. By

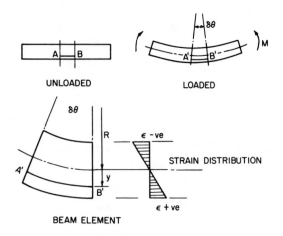

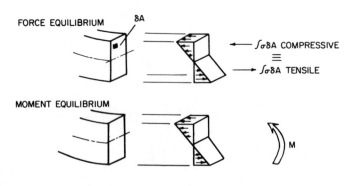

Figure 4.6

dealing with a constant bending moment, relationships are derived which can be used for a moment which varies along the length. The cross-sectional shape is taken to be rectangular for simplicity but the simple bending theory developed here is valid for any section which is symmetrical about an axis in the plane of bending. Sections unsymmetrical to the plane of bending are dealt with in a similar but more complicated manner.

Conventionally it is usual to start from compatibility and examine possible deformation states for the beam. If each elemental length is to deform in the same way it is usually taken that 'plane sections remain plane', i.e. initially straight sections normal to the axis of the beam remain straight after deformation, for example the ends of the beam remain flat. This seems plausible from a compatibility point of view since, because of symmetry, any other choice such as curved deformed planes would require gaps in the beam. Thus the only possible compatible deformed shape is a circular arc. These deductions are confirmed by experiment. Since one face is obviously compressed and the other stretched it will be readily accepted that there will be a neutral surface that does not change its length. A transverse axis of the beam lying in this neutral plane is termed the neutral axis. The radius of curvature of this neutral axis is taken as R although its location is not yet known. The strain at any fibre is

$$\varepsilon = \frac{A'B' - AB}{AB} = \frac{(R+y)\delta\theta - R\delta\theta}{R\delta\theta} = y/R \qquad (4.3)$$

From equation (4.3) it is clear that the strain is proportional to the distance from the neutral axis and thus has a linear distribution. The position of *zero* strain is not yet known. The linear elastic stress–strain relation ($E = \sigma/\varepsilon$) is now introduced and it is found that

$$\frac{\sigma}{y} = \frac{E}{R} \qquad (4.4)$$

showing that the stress distribution is also linear.

It should be noted, as will become clear later, that the strain in one direction is usually dependent on the stress in all (three) directions. However, for the case of small cross-sectional dimensions compared with length the two transverse stresses are assumed zero since there are no externally applied forces in these directions. This would automatically exclude 'plates' or 'wide beams', etc.

Finally, equilibrium is considered. Two conditions relevant to the longitudinal stress system of equation (4.4) are available. The longitudinal force is zero and the external moment must be balanced.

Thus

(1)
$$\int dF = 0$$

i.e.
$$\int \sigma \, dA = 0$$

or
$$\int \frac{E}{R} y \, dA = 0$$

Since E and R are simply constants for a given loaded beam and y is the distance from the neutral axis, the above statement indicates that the first moment of area around the neutral axis is zero. Since this is only true if y is measured from an axis passing through the centroid then it follows that the neutral axis must pass through the centroid of the cross section,

(2)
$$\int \sigma y \, dA = M$$

$$\int \frac{E}{R} y^2 \, dA = M \tag{4.5}$$

The term $\int y^2 \, dA$ is the second moment of area of the cross section around the neutral axis, usually denoted by I, and sometimes misleadingly called the moment of inertia. Hence

$$\frac{M}{I} = \frac{E}{R} \tag{4.6}$$

Combining (4.4) and (4.6) gives the well known simple beam bending equation

$$\frac{M}{I} = \frac{\sigma}{y} = \frac{E}{R} \tag{4.7}$$

For a rectangular section b wide by d deep, $I = bd^3/12$, and for a hollow circular section of external and internal diameters D and d, $I = (\pi/64)(D^4 - d^4)$, etc.

When calculating the second moment of area of a complicated shape, such as a group of weld runs, the *parallel axes theorem* is extremely useful. Considering *Figure 4.7*, the second moment of area of the shaded shape round the x and y axes is given by

$$I_{xx} = I_{aa} + Ah^2$$
$$I_{yy} = I_{bb} + Aj^2$$

Convenient simplifications can often be made; in this case for example I_{aa} will be small compared with Ah^2, and could be neglected.

Since the beam equation has utilised all three of the basic principles and is used so widely in problem solving, it is sometimes difficult to keep track of the basic steps in even the simplest of problems.

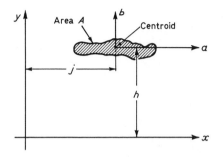

Figure 4.7

The compatibility equation (4.3) and the equilibrium equation (4.5) are obviously independent of the material behaviour. It is instructive to repeat the steps for a different stress–strain law. For example, $\varepsilon = B\sigma^n$ is a well-known and useful expression to describe a nonlinear elastic material where B and n are constants and n is an odd number for convenience. (n odd ensures that negative stress results in negative strain without further mathematical complexity.) The resulting beam equation is

$$\frac{2^{(n+1)}\left(\dfrac{2n+1}{n}\right)^n M^n}{b^n d^{2n+1}} = \frac{1}{BR} = \frac{\sigma^n}{y}$$

Note that it is not easy to define an I value which has a physical meaning although a 'modified moment of inertia' could be defined if desired. When $n = 1$ the above equation is identical to equation (4.7). The longitudinal force equilibrium condition that in the linear case indicated the neutral axis coincided with the centroid is not easy to solve in the nonlinear case. The neutral axis does happen to coincide with the centroid for the symmetrical section considered here but will not in general do so for nonlinear material behaviour.

It is perhaps worth pointing out in passing that the bent beam is a kinematically determinate structure in the sense that if one knows or measures the strain anywhere then the strain can be obtained anywhere else from compatibility alone, i.e. equation (4.3). On the other hand the beam is statically indeterminate in the sense that the stress system cannot be determined from a knowledge of the applied moment using equilibrium alone. It is necessary to utilise the deformation behaviour as in the above derivation.

Torsion action

Torsion action can similarly be used to illustrate the basic principles. Consider the circular cross-section shaft shown under the action of a torque T (*Figure 4.8*). As in the bending case, compatibility is

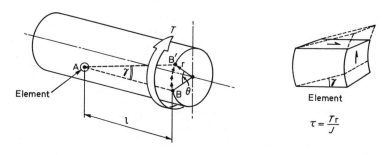

$$\tau = \frac{Tr}{J}$$

Figure 4.8

considered first by studying the possible geometry of deformation. A fibre AB on the unloaded shaft will become AB' on the loaded shaft. The shear strain defined as the change in angle of any right angle in a rectangular element is given by the angle γ (BAB'). The shear strain is then

$$\gamma = \frac{r\theta}{l} \tag{4.8}$$

A linear elastic constitutive relationship for shear relates the shear stress τ to the shear strain γ via the modulus of rigidity G.

$$\tau = G\gamma \tag{4.9}$$

Applying equation (4.9) to equation (4.8) gives

$$\frac{\tau}{r} = \frac{G\theta}{l} \tag{4.10}$$

which is generally known as the first torsional relationship.

Hence from equation (4.10) it can easily be seen that for a given shaft experiencing a given twist, the shear stress is linearly proportional to the radius, and is a maximum at maximum radius.

Equilibrium is now considered by taking moment equilibrium about the centre of the shaft.

$$T = \int (\text{shear force on an element}) \times \text{radius}$$

Consider a circular element at radius r, thickness dr:

$$T = \int \tau 2\pi r^2 \, dr$$

Substituting for τ from equation (4.10), integrating from the centre to the outside of the shaft and rearranging gives

$$\frac{T}{J} = \frac{G\theta}{l} \tag{4.11}$$

where equation (4.11) is known as the second torsional relationship and $J = \pi r^4/2$ is a property of the cross section and is sometimes referred to as the Polar Moment of Inertia; r here is the outside radius of the shaft. In the case of a hollow shaft

$$J = \frac{\pi}{2}(r_0^4 - r_i^4)$$

Combining (4.10) and (4.11) gives the torsional result analogous to the bending equation (4.7).

$$\frac{T}{J} = \frac{\tau}{r} = \frac{G\theta}{l} \tag{4.12}$$

Notice that both the shear strain and the shear stress vary from zero at the centre to maximum at the outside of the shaft. For those not familiar with the idea of polar moment of inertia, consider a set of orthogonal axes (x, y, z) at the centre of the circular cross section, x and y being in the plane and z normal to the plane. Then

$$J = I_{zz} = I_{xx} + I_{yy} \quad (\textit{perpendicular axes theorem})$$

where I is the second moment of area about an axis in the plane, as used for bending. For the circular section, $I_{xx} = I_{yy}$, $I = \pi r^4/4$, and $J = 2I$.

Again linear elastic material has been assumed but the reader may find it helpful to repeat the steps for the nonlinear law already mentioned ($\gamma = C\tau^n$) whereupon the result becomes

$$\frac{T}{J_c} = \frac{\tau}{r^{1/n}} = \left(\frac{1}{C}\frac{\theta}{L}\right)^{1/n}$$

where

$$J_c = \frac{2n}{3n+1} \pi r^{(3n+1)/n}$$

4.1.3 Pressurised cylinders

Previous examples have been limited to essentially uniaxial problems, i.e. stresses and strains need only be considered in one direction at a time. Welds are frequently located at points of geometric discontinuity where multiaxial rather than uniaxial analyses are appropriate, and it is therefore useful to give an example of such an analysis. In the case of a cylinder under internal (or external) pressure, we ought to consider stress and strain in three directions—circumferentially, longitudinally and radially. However it turns out that a simple approximate solution can be developed if the cylinder is reasonably thin, and the radial stress is small compared with the others.

*Thin cylinders (**Figure** 4.9)*

Considering the axial symmetry of the problem, the stresses should not vary round the circumference and there should be no shear stress parallel to the cylinder wall. The hoop stress σ_θ could vary across the wall thickness (clearly the radial stress varies from $-p$ to zero).

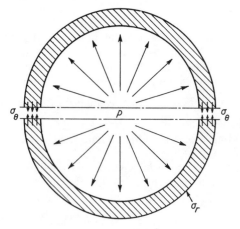

Figure 4.9

The forces developed by σ_θ in the cylinder wall at any cross section must be in equilibrium with the component of pressure force normal to that section. Hence the *average* hoop stress is given by $\sigma_\theta = pd/2t$ where d is the diameter. The longitudinal stress is given also by equilibrium as

$$\sigma_l = \frac{pd}{4t}$$

The hoop strain can then be found via the appropriate two-dimensional stress/strain relation, and the increase in radius u from compatibility (simple geometry) i.e.

$$\frac{u}{r} = \varepsilon_\theta$$

Notice that the stress was established from equilibrium alone, and therefore the solution is valid for any material (steel, rubber, soap, etc.). The snag is that we have no idea of the variation of σ_θ and σ_l from the average values—they may be much higher or lower at different radii. This important point cannot unfortunately be resolved without considerable extra effort. Although it is a little beyond the scope of this book the complete solution will now be detailed to show how far one needs to go to be satisfied with the answer even in the case of a simple component. In passing it should be noticed that a similar result is obtained for a pressurised thin sphere, namely, $\sigma_\theta = pd/4t$ which happens to be the same as the longitudinal stress in a pressurised cylinder.

Thick cylinders

A thick cylinder under internal pressure p is shown schematically in *Figure 4.10*. Consider an element at some position part way through the thickness of the cylinder wall. The possibility of stress variation over the element is introduced, i.e. σ_r and σ_θ are functions of the radius. By arguments of symmetry it can be deduced that there are no shear stresses and that the stresses will be those shown.

Equilibrium of forces in the radial direction gives

$$(\sigma_r + \delta\sigma_r)(r + \delta r)\delta\theta\delta l = \sigma_r r\delta\theta\delta l + \sigma_\theta \delta r\delta l\delta\theta$$

which, neglecting small quantities, is

$$r\,\mathrm{d}\sigma_r + \sigma_r\,\mathrm{d}r = \sigma_\theta\,\mathrm{d}r$$

and combining terms gives

$$\sigma_\theta = \frac{\mathrm{d}}{\mathrm{d}r}(r\sigma_r) \tag{4.13}$$

From compatibility it can easily be seen that the longitudinal strain cannot vary with radius, otherwise barrelling or bulging would result.

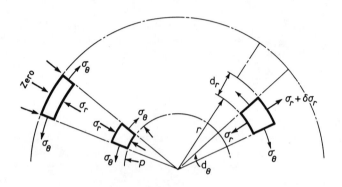

Figure 4.10

The circumferential and radial strains do not vary in the circumferential direction from symmetry. Hence

$$\varepsilon_\theta = \frac{\text{change in circumference}}{\text{original circumference}} = \frac{2\pi(r+u) - 2\pi r}{2\pi r} = \frac{u}{r}$$

where u is the radial displacement at any radius r, and

$$\varepsilon_r = \frac{\text{change in thickness}}{\text{original thickness}} = \frac{\delta u}{\delta r} = \frac{\mathrm{d}u}{\mathrm{d}r} \quad \text{in the limit}$$

Combining the two strain conditions gives a single compatibility equation.

$$\varepsilon_r = \frac{\mathrm{d}}{\mathrm{d}r}(r\varepsilon_\theta) \tag{4.14}$$

In the present problem where stresses occur in more than one direction, the constitutive relationship is more complicated. This will be discussed later in this section. For linear elasticity the relationships are

$$\varepsilon_\theta = \frac{1}{E}\left[\sigma_\theta - v(\sigma_r + \sigma_l)\right]$$

$$\varepsilon_r = \frac{1}{E}\left[\sigma_r - v(\sigma_\theta + \sigma_l)\right]$$

$$\varepsilon_l = \frac{1}{E}\left[\sigma_l - v(\sigma_\theta + \sigma_r)\right] \qquad (4.15)$$

where v is Poisson's ratio.

σ_l may be eliminated from the above equations which reduces them to

$$E\varepsilon_r = \sigma_r(1 - v^2) - v(1 + v)\sigma_\theta - vE\varepsilon_l$$
$$E\varepsilon_\theta = \sigma_\theta(1 - v^2) - v(1 + v)\sigma_r - vE\varepsilon_l$$

These may now be substituted into the compatibility equation (4.14) and using equation (4.13) gives, after some manipulation, an equation in σ_r only.

$$\sigma_r = \frac{d}{dr}r\frac{d}{dr}(r\sigma_r) \qquad (4.16)$$

Equation (4.16) is a differential equation which has a solution of the form

$$\sigma_r = A + B/r^2 \qquad (4.17)$$

which can be checked by substitution. Substitution back into (4.13) gives the circumferential stress

$$\sigma_\theta = A - B/r^2 \qquad (4.18)$$

Equations (4.17) and (4.18) are the well-known Lamé equations for a thick cylinder. A and B are constants which may be found from the boundary conditions. It is usually convenient to leave the Lamé equations in this general form rather than specify these constants for a particular case of, say, internal pressure p, external pressure zero.

It will be seen that $\sigma_\theta + \sigma_r = 2A = $ constant and since the longitudinal strain is also constant, the stress–strain relationships indicate that σ_l will be constant (equation (4.15)) and can easily be found from equilibrium consideration in the longitudinal direction. For the case of internal pressure p alone, if the internal and external radii are *a* and *b* respectively, then

$$\sigma_l = \frac{\text{net pressure end load}}{\text{cross-sectional area}} = \frac{pa^2}{(b^2 - a^2)}$$

and from equations (4.17) and (4.18) the other stresses become

$$\sigma_r = \frac{pa^2}{(b^2 - a^2)}\left(1 - \frac{b^2}{r^2}\right)$$

$$\sigma_\theta = \frac{pa^2}{(b^2 - a^2)}\left(1 + \frac{b^2}{r^2}\right) \tag{4.19}$$

The solution is, of course, like the others, only valid away from disturbances caused by end connections, holes, etc.

Now that equation (4.19) has been established it is relatively easy to check the simple equilibrium solution for σ_θ. It is not difficult to show that if $d > 20t$ the error in $\sigma_\theta = pd/2t$ compared with equation (4.19) is about 5% or less. Hence the connotation 'thin' cylinder.

4.1.4 Shear action

A component supported in some unspecified manner and loaded as shown in *Figure 4.11*, will be subjected simultaneously to bending and shear; on the face of it a complex situation. It is simplified by assuming that the bending deformation of an element within the

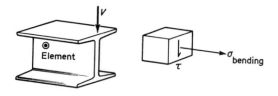

Figure 4.11

beam does not affect the shear stresses and vice versa. Thus, by turning a blind eye to one of the compatibility requirements, a simpler, though less exact solution for the shear stresses is obtained by considering equilibrium alone. Like the previous thick-cylinder example, the full solution can be used to show retrospectively that the approximate solution given here is indeed exact for a constant shear force along the beam.

The transverse loading case giving rise to combined bending and shear is, of course, a more realistic means of loading than the pure bending case previously discussed. However it is usually agreed that the bending stresses as evaluated from equation (4.7) are accurate enough for the combined case except of course that the bending moment usually varies along the length of the beam and the longitudinal stress varies accordingly. The shear stress situation will now be considered separately.

A full discussion is beyond the scope of this section (see references 1 or 2 for details). The restrictions are similar to those for the bending alone case. The shear stress at any distance from the neutral axis can be derived as (see *Figure 4.12*)

$$\tau = \frac{VA\bar{y}}{bI} \qquad (4.20)$$

where A is the shaded area above the level being considered, $\bar{y}$ is the distance from the neutral axis for bending to the centroid of the area and b the width at the level of interest.

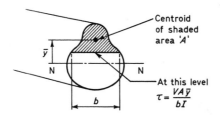

Figure 4.12

Equation (4.20) may be applied to common shapes of cross section. For example a rectangular section of dimensions $b \times d$. Some detailed consideration of equation (4.20) will show that $\bar{y}$ is zero for the top and bottom surfaces of the beam. The shear stress distribution varies from zero in parabolic fashion through a maximum at the centre and back to zero. At the centre it is

$$\tau = \frac{3}{2}\frac{V}{A}$$

Notice that the average shear stress obtained by assuming it to be distributed uniformly everywhere across the section is simply load/area which is V/A. In a similar way maximum shear stresses can be found for a circular bar and a thin tube (see *Figure 4.13*). They are approximately

$$\tau = \frac{4}{3}\frac{V}{A}$$

and

$$\tau = 2\frac{V}{A}$$

In the above three examples A is the total area of cross section.

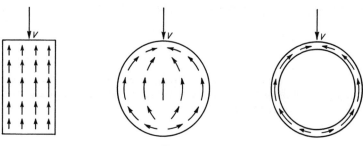

Figure 4.13

It is worth noting that the flow of shear stress across the section is such that it is parallel with the free surfaces. This arises because there can be no shear stress normal to the free surface because of the condition of complementary shear which will be discussed later under the heading 'Combined Loading Systems'.

An example of particular relevance to engineering structure is the I beam (see *Figure 4.14*) which may be assumed to be made up of

Figure 4.14   *I beam*

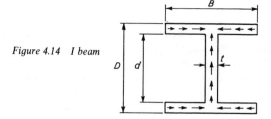

rectangular portions. The I section is a good shape for giving resistance to bending in the plane of the section. Material is located (in the flanges) where bending stresses are high and shear effects are largely carried by the web. The distribution of shear in the web is again parabolic being maximum in the centre. Despite the fact that values can be obtained from equation (4.20), vertical shear in the flanges is usually taken to be negligible. The stress distribution at the junction of the web and flange is rather complicated and in rolled steel beams generous fillet radii are provided to minimise the stress concentration effect. In the flanges the horizontal shear stresses are significant and they vary linearly from zero at the edges to a maximum at the centre. For many common I beams the maximum shear stress at the centre of the web can be approximated by

$$\tau = \frac{V}{A}$$

In most of the above, a proper analysis of shear effects has been glossed over since space precludes more detailed treatment. This is probably a fair representation of many engineers' knowledge of the subject. Unfortunately, however, the welding design engineer often finds his welds require to be assessed for shear effects (e.g. a fabricated I beam) and a fuller understanding would be most advantageous. One of the most common traps in the design of fabricated sections to resist bending is to assume that the simple beam bending formula given earlier *always* works, regardless of section shape and loading. In welded construction it is altogether too easy to design a section with wide flanges (relative to the web depth) in a short span, where there is little or no space to feed the bending loads into the extremities of the flanges (see *Figure 4.15*). In these circumstances, parts of the flanges strain rather less than the simple theory imagines (i.e. plane sections become crooked!), the stiffness of the section decreases and the stresses in the parts of the flanges which are still working, increase. This is the 'shear lag' phenomenon which is treated

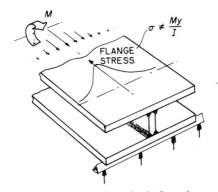

Figure 4.15 Bending of wide-flange beams

in reference 4 along with many other interesting facets of non-standard beam behaviour.

At this stage something must be said about beams which are not symmetrical in the plane of bending. An increasing number of beam sections are being used which fit the category of having the thickness small in comparison with other geometric parameters and having only one or no axis of symmetry. These are broadly described as 'thin-walled open sections'. Perhaps the most familiar is the channel section (see *Figure 4.16a*). Generally for these shapes if external shear forces act through the centroid of the cross section then twisting as well as bending will result. To avoid twisting the transverse loads must be located such that they act through a position which is known as the 'shear centre'. The location of the shear centre is dependent only on the geometry of the cross section and the direction of loading.

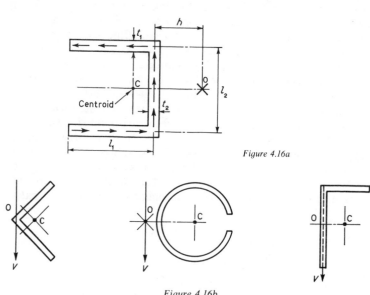

Figure 4.16a

Figure 4.16b

For example in a channel section subject to a vertical shear the shear stress flow will be as indicated. Vertical equilibrium dictates that the sum of the vertical shear flow in the web must balance the applied force V. Horizontal equilibrium dictates that the shear forces in the top and bottom flanges must be equal in magnitude but opposite in direction. If V were to act through C it will be evident that V and the vertical shear in the web produce an anti-clockwise twisting moment. So do the forces in the flanges. Even if

V is located to act through the web the flange forces still cause a twisting moment. It becomes evident that V must act through some point outside of the section distance h away to avoid twisting. This point O is called the shear centre. For the channel section

$$h = \frac{3l_1^2 t_1}{l_2 t_2 + 6l_1 t_1}$$

Good practice in applying loads to channel sections obviously requires some bracket attachment to avoid twisting. It is relevant to emphasise that the previous simple beam bending theory is only valid in the absence of twisting. A few other cases of shear centre are shown diagrammatically for interest in *Figure 4.16b*.

A word of caution in connection with shear centre. What has been said above is true for linear elastic material response. In fact the location of the shear centre depends on the material behaviour. While unlikely to be a major consideration in design of common structures it should be noted. For example for fully plastic conditions (see *Figure 4.3* and Section 4.3) the shear centre of the channel section is located on the web and not outside the section.

In discussing shear action one must look again at torsion, this time on noncircular cross sections. For instance consider the *thin* variable thickness closed tube subject to a torque T (see *Figure 4.17*). Briefly, the shear stress in the cross section is taken to be parallel to the boundary and the shear force per unit length is q. For equilibrium of any element with its neighbours this force must be constant so that the shear stress at any section can be found from $\tau = q/t$ where t is the

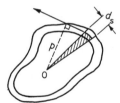

Figure 4.17

local thickness. (The shear stress is also assumed uniform across the wall thickness because of the 'thin' condition specified.) Taking moments about some point O within the tube

$$T = \int_0^{2\pi} qp \, ds$$

where p is the perpendicular distance shown, or

$$T = 2q \int_A dA$$

where dA is the shaded area. Thus

$$T = 2qA$$

where A is the area enclosed by the middle surface of the tube wall. Hence

$$\tau_{max} = \frac{T}{2At_{min}}$$

Again the approximate solution above has been derived from equilibrium considerations only by making a simplifying guess as to the nature of the main stress and neglecting others. An idea of how approximate the analysis might be can be obtained by comparing it with the exact solution for the thin-walled circular tube of constant t as obtained from equation (4.12). Reference 2 shows that for a radius ratio of 0.9 there is a 4.7% difference and for a radius ratio of 0.75 the difference is 10.7%.

Torsion of a narrow rectangular strip of thickness t and depth d can be found by approximating it by a thin flat tube whereupon the maximum shear stress can be found as

$$\tau_{max} = \frac{3T}{dt^2}$$

Channel and I beam sections can then be treated as assemblies of rectangular cross sections.

Summary

Figure 4.18 summarises a number of results for simple loading systems in a form suitable for general use. It should be remembered that most of the solutions are inexact, and often contain restrictions on the dimensions of the section which can be treated without significant error (e.g. thin cylinder formula not suitable for $d/t < 20$).

Reference 4 is a useful handbook which gives many stress-analysis solutions in a condensed form.

4.1.5 Combined loading systems

Unfortunately for the designer, real components are very rarely loaded in a simple fashion. If one thinks of a marine propeller shaft, it may experience direct load, torsion, bending and shear simultaneously (*Figure 4.19*).

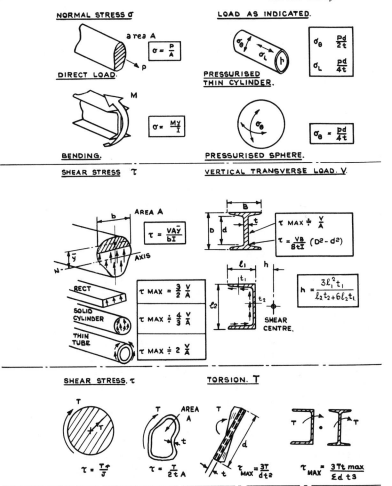

Figure 4.18 Summary of results for simple loading systems

At any welded joint in a component subject to combined loading, stresses will arise from the different load actions previously described and they will obviously vary from place to place. It is useful at this stage to introduce the concept of 'stress at a point'. A small cubic element of material is considered at a point, with its sides parallel to known coordinate directions (*Figure 4.20*). If the element is small the stresses may be considered constant across its faces and represented as shown. It would seem that there are three possible values of normal stress and six values of shear stress. However, by moment

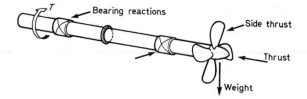

Figure 4.19

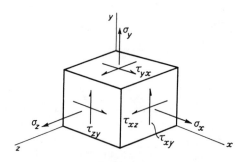

Figure 4.20

equilibrium of the cube about each of the coordinate axes, the shear stresses are reduced to three. It is worth noting that it is impossible to apply a shear stress to one face of an element without generating stresses (through equilibrium) on the other three faces. The stresses on faces at right angles to the 'active' faces are sometimes called 'complementary shear stresses'.

The three-dimensional situation shown here is reduced wherever possible to a two-dimensional system, where the stress on one of the planes (e.g. the z plane) is zero or smaller by comparison. This situation ($\sigma_z = 0$) is known as *plane stress* and gives a simpler two-dimensional element (*Figure 4.21*). If, for example, a shaft is

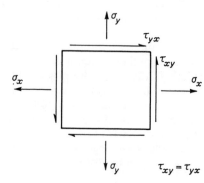

Figure 4.21

$\tau_{xy} = \tau_{yx}$

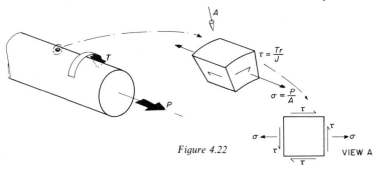

Figure 4.22

VIEW A

subject to torsion T and direct load P as shown in *Figure 4.22*, any element at or beneath the surface can be reduced to the two-dimensional element shown.

So far the elements chosen (representing points) have been conveniently orientated in order that the relevant stresses could be easily calculated according to some 'simple system' theory. This is not always possible or desirable and, in any case, it is instructive to consider the stress state on any plane inclined at some arbitrary angle. It would be convenient to have a means of transforming the stresses from one plane to another.

4.1.6 Stress transformations (for plane stress)

Consider the element shown in *Figure 4.23* with an inclined plane at an angle θ. By applying force equilibrium conditions to the shaded portion, the following two equations can be derived:

$$\sigma_\theta = \sigma_x \cos^2 \theta + \sigma_y \sin^2 \theta + \tau_{xy} \sin 2\theta$$

$$\tau_\theta = \frac{\sigma_y - \sigma_x}{2} \sin 2\theta + \tau_{xy} \cos 2\theta \qquad (4.21)$$

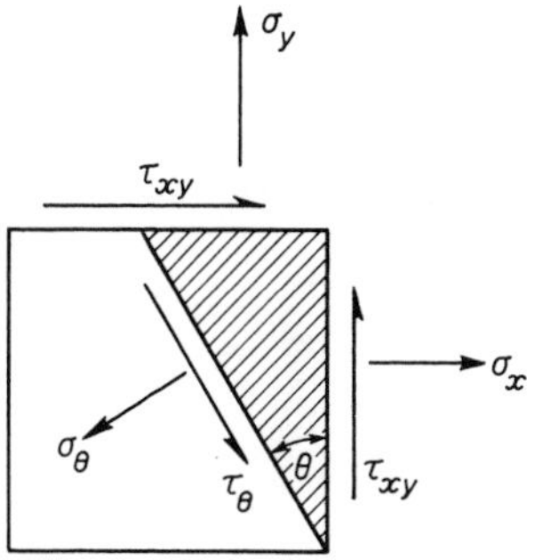

Figure 4.23

A reader who is not familiar with these equations may benefit by proving them for himself. Note that stresses are not in equilibrium but forces (stress × area) are.

Solution of the stress transformation equations is lengthy and tedious (although they may easily be programmed for a computer) and in about 1882, Mohr noticed that if they were manipulated they could be arranged in the form of the equation of a circle (see *Figure 4.24*).

$$\left(\sigma_\theta - \frac{\sigma_x + \sigma_y}{2}\right)^2 + \tau_\theta^2 = \left(\frac{\sigma_x - \sigma_y}{2}\right)^2 + \tau_{xy}^2 \qquad (4.22)$$

The circle drawn in a (σ, τ) coordinate system has its centre on the σ axis at the point

$$\left(\frac{\sigma_x + \sigma_y}{2}\right), 0$$

and its radius is given by

$$\left[\frac{(\sigma_x - \sigma_y)^2}{2} + \tau_{xy}^2\right]^{\frac{1}{2}}$$

Important features of the Mohr circle graphical solution described are as follows:

1 The completed circle defines the state of two-dimensional stress existing *at a point* in a stressed component.
2 A point on the circle represents a plane in the material, at the point under consideration.
3 The normal stress on the plane is given by the σ coordinate of the point, both in magnitude and sign. (Tensile is positive.)

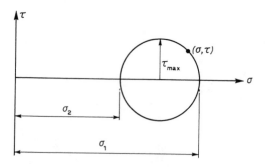

Figure 4.24

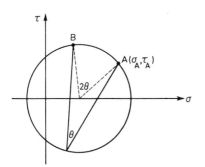 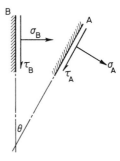

Figure 4.25

4 The shear stress on the plane is given by the τ coordinate of the point. The direction of the shear stress (i.e. whether it tends to rotate the element clockwise or anticlockwise) is given by the sign of the τ coordinate. (Clockwise positive usually.)

5 The angle between two planes in the material is given by half the angle subtended by the points representing the planes at the centre of the circle, or alternatively by the *angle* subtended at the circumference of the circle (see *Figure 4.25*).

6 Maximum and minimum values of normal stress σ_1, σ_2 (called *principal stresses*) occur on mutually perpendicular planes (called *principal planes*), and on these planes the shear stress is zero.

7 The maximum values of shear stress occur on planes inclined at $45°$ to the principal planes, and $\tau_{max} = (\sigma_1 - \sigma_2)/2$.

Note

Throughout the foregoing analysis of stress transformations, the only principle which has been used is the principle of equilibrium. Material properties or deformations have not been mentioned. The analysis, therefore, applies without exception to all materials. The only restriction is that the material must not deform too much under load so that the change in geometry of the component influences the equilibrium conditions assumed for the undeformed shape. An obvious example is provided by the pressurised sphere represented by a child's balloon. Earlier the stress in such a component was derived as $pd/4t$ where d and t were diameter and thickness.

However it will be obvious that if the diameter increases and the thickness decreases, as in an inflating balloon, the stress as calculated from the original values has little relevance. Fortunately in most engineering applications the changes in dimensions are usually small compared with their original values.

Example 4.1

A thin square section fabricated and stress-relieved steel tube is loaded by fluctuating torque T, and moment M which are in a constant ratio. A fatigue crack develops on a tensile principal plane as shown in *Figure 4.26*. Find the ratio T/M.

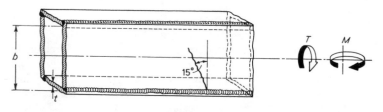

Figure 4.26

Solution

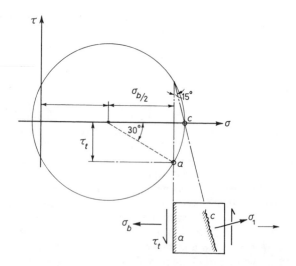

Figure 4.27

From the Mohr circle (*Figure 4.27*)

$$\frac{\tau_t}{\sigma_b/2} = \tan 30°$$

hence

$$\tau_t = \frac{\sigma_b}{2\sqrt{3}} = \frac{My_{max}}{T2\sqrt{3}} = \frac{T}{2tA}$$

Therefore

$$\frac{T}{2tA} \simeq \frac{Mb/2}{2(\frac{1}{12}b^3t + \frac{1}{4}b^3t)}\frac{1}{2\sqrt{3}} \text{ (term neglected in } I \text{ is } 2 \times \frac{1}{12}t^3b)$$

and

$$\frac{T}{M} = \frac{2b^2t \times b/2}{2b^3t(\frac{1}{12}+\frac{1}{4})}\frac{1}{2\sqrt{3}}$$

$$\frac{T}{M} = 0.434$$

and the direction of T is opposite to that originally shown on the sketch of the tube.

Superposition

If two different equilibrium stress systems are active at a point in a material, but on different planes, the complete state of stress at the point may be found by transforming the stresses from each system onto common planes, and adding them algebraically. This may be useful where service stresses are added to a set of residual stresses. In such a case, superposition is only valid as long as the material is linear (i.e. below yield). If the two sets of stresses are individually statically determinate with respect to the loads causing the stresses, then the stresses may be superposed even if the material is nonlinear (e.g. in a thin tube subject to torque and internal pressure the individual stress systems in *Figure 4.18* may be superposed).

Example 4.2

If in a 45° spiral welded tube the residual principal stresses of 200 and 50 MN/m^2 are as shown in *Figure 4.28*, find the combined stress

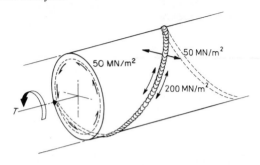

Figure 4.28

state if a torque which generated 50 MN/m² shear stress on a transverse cross section is applied to the tube.

Solution (Figure 4.29)

The torsion shear stress transformed to 45° planes gives + 50 MN/m² normal stress transverse to the weld line, and − 50 MN/m² parallel to the line. Hence the total principal stresses are 150 MN/m² parallel, and 100 MN/m² transverse to the weld line. If the torque direction is reversed, the stresses are 250 MN/m² and zero (provided that the yield strength is 250 MN/m² or greater).

4.1.7 Strain transformations (two dimensional)

In order to predict the deformations of engineering components a knowledge of strain is needed, as well as stress. A state of three-dimensional strain at a point in a deforming body can always be reduced to a combination of normal strain in three directions (giving change in volume) and shear strain in three directions (giving change in shape). The variation in strain from point to point, and the variation from plane to plane at a point, are controlled by pure geometry (compatibility) in a homogeneous body. In *Figure 4.30* a two-dimensional strain is applied to a square element.

The strain normal to the plane AX may be calculated by measuring the extension of the line BP to BP'. Similarly, by measuring the change of angle AOP the shear strain on the plane AX can be determined. Then from pure geometry it can be deduced (eventually)

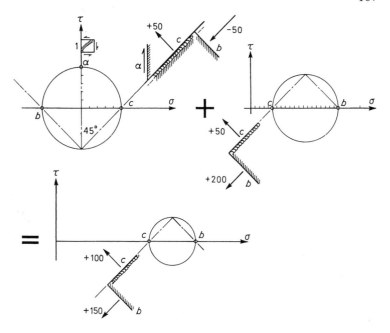

Figure 4.29

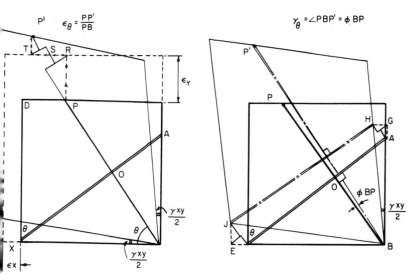

Figure 4.30   Two-dimensional strain applied to a square element

that

$$\varepsilon_\theta = \varepsilon_x \cos^2 \theta + \varepsilon_y \sin^2 \theta + \frac{\gamma_{xy}}{2} \sin 2\theta$$

$$\frac{\gamma_\theta}{2} = \frac{\varepsilon_y - \varepsilon_x}{2} \sin 2\theta + \frac{\gamma_{xy}}{2} \cos 2\theta \qquad (4.23)$$

It is not suggested that these should be derived without reference to other books. It is not difficult but the definitions of strains are important.

Again, in a similar way as for the stresses, equation (4.23) can be manipulated into the equation of a circle:

$$\left(\varepsilon_\theta - \frac{(\varepsilon_x + \varepsilon_y)}{2}\right)^2 + \left(\frac{\gamma_\theta}{2}\right)^2 = \left(\frac{\varepsilon_x - \varepsilon_y}{2}\right)^2 + \left(\frac{\gamma_{xy}}{2}\right)^2 \qquad (4.24)$$

Comparison of equations (4.23), (4.24) with (4.21), (4.22) shows that they are identical provided that ε_θ is taken as equivalent to σ_θ and $\gamma_\theta/2$ equivalent to τ_θ.

Hence a Mohr circle solution may be used for strain problems. The statements made about the Mohr circle for stress all apply to the Mohr circle for strain.

Again it should be noted that when the strain transformation equations are used on their own, they apply without restriction to any material no matter how it is strained, whether by stress, thermal expansion, chemical changes, etc. The only restriction is that the strain as used here is the usual engineering strain defined relative to the undeformed original geometry. For large deformations such a definition may not be very meaningful. Large in this context means large enough to influence the geometric calculations in deriving the equations.

4.1.8 Stress–strain relations in two dimensions

The link between the strains experienced by an engineering component and the stresses applied to it arise through the above relations. Engineers rely on mechanical tests on representative material, to determine these properties. Here attention shall be confined to linear elastic isotropic behaviour.

Normal stress

In one dimension, $\varepsilon = \sigma/E$. However, if the stress system is multiaxial, Poisson's ratio interferes with the simplicity of this relationship.

Poisson's ratio is defined as the ratio of the strain in a direction perpendicular to the load, to the strain in the direction of the load.

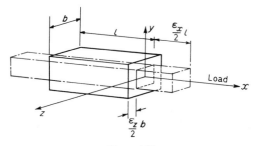

Figure 4.31

In the example shown (*Figure 4.31*) the strains in the y and z directions are

$$\varepsilon_y = \varepsilon_z = -\nu\varepsilon_x$$

Poisson's ratio ν is a physical property. Under classic conditions it varies between 0.25 and 0.35 for common engineering materials. For steel, $\nu = 0.28$. From geometrical considerations, Poisson's ratio for constant-volume deformation (such as in plasticity) is 0.5. The apparent value of ν is also greatly influenced by anisotropy.

The complete relationships for three-dimensional stress and strain are:

$$\varepsilon_x = \frac{1}{E}\left[\sigma_x - \nu(\sigma_y + \sigma_z)\right]$$

$$\varepsilon_y = \frac{1}{E}\left[\sigma_y - \nu(\sigma_x + \sigma_z)\right]$$

$$\varepsilon_z = \frac{1}{E}\left[\sigma_z - \nu(\sigma_x + \sigma_y)\right]$$

If, in addition to strain due to stress, there is also a thermal strain due to a change in temperature then these relationships become typically

$$\varepsilon_x = \frac{1}{E}\left[\sigma_x - \nu(\sigma_y + \sigma_z)\right] + \alpha\Delta t$$

where α is the linear coefficient of expansion and Δt the change in temperature.

Shear stress

Shear strain involves only change of shape and is unaffected by Poisson's-ratio effects, or change in temperature unless in specially contrived situations. Hence:

$$\gamma_{xy} = \frac{\tau_{xy}}{G}$$

$$\gamma_{yz} = \frac{\tau_{yz}}{G}$$

$$\gamma_{xz} = \frac{\tau_{xz}}{G}$$

where G is the modulus of rigidity.

The modulus of rigidity is related to Young's modulus by

$$G = \frac{E}{2(1+v)}$$

If $\sigma_z = 0$ (plane stress) the equation can be easily rearranged to give the normal stresses directly:

$$\sigma_x = \frac{E}{1-v^2}(\varepsilon_x + v\varepsilon_y)$$

$$\sigma_y = \frac{E}{1-v^2}(\varepsilon_y + v\varepsilon_x) \tag{4.25}$$

If, on the other hand, $\varepsilon_z = 0$ as in plane strain,

$$\sigma_z = v(\sigma_x + \sigma_y)$$

and the general equations can be written

$$\varepsilon_x = \frac{1-v^2}{E}\left(\sigma_x - \frac{v}{1-v}\sigma_y\right)$$

$$\varepsilon_y = \frac{1-v^2}{E}\left(\sigma_y - \frac{v}{1-v}\sigma_x\right)$$

Stresses determined from strain measurements (application of two-dimensional stress and strain transformations)

As discussed in Chapter 3, it will not always be possible to calculate stresses directly from the loads and the structural geometry; perhaps because the loads arise from natural or uncontrolled forces, or because the geometric shape does not lend itself to easy structural analysis. In such cases it may be helpful to make strain measurements on the real structure in service, and infer the stresses via appropriate stress–strain relations for the material. Electrical resistance strain gauges are commonly used for this kind of work, and reference 6 contains a useful survey of current techniques.

Example 4.3

A 60° array of strain gauges is bonded to a steel component which forms part of an earthmoving machine. While the machine is working, the following peak strain readings are measured simultaneously:

$$A = +700 \text{ microstrain (i.e. } \times 10^{-6})$$
$$B = -135 \text{ microstrain}$$
$$C = -402 \text{ microstrain}$$

Find the directions and magnitudes of principal stresses relative to the weld line. (Assume $E = 200 \text{ GN/m}^2$)

Solution

From *Figure 4.32*

$$\varepsilon_1 = +718 \text{ microstrain}$$
$$\varepsilon_2 = -610 \text{ microstrain}$$

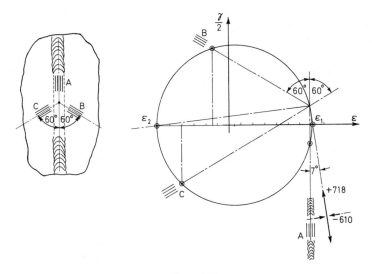

Figure 4.32

from equation (4.25)

$$\sigma_1 = \frac{200 \times 10^9}{(1 - 0.28^2)} (718 - 0.28 \times 610) \times 10^{-6}$$

$$= +119 \text{ MN/m}^2$$

$$\sigma_2 = -89 \text{ MN/m}^2$$

tensile principal stress is offset from the weld line by 7°.

4.1.9 Buckling

When certain structures are loaded, they change their shape or dimensions in such a way that they become less able to carry further increase in load. This can be dangerous, as a small increase in load may lead to a large deformation, even if the material remains within its elastic limit. The toy balloon previously mentioned is an example of just such a load–shape instability, but in practical engineering situations instability more often occurs through buckling, where compression loading is applied to a shape which is thin relative to the length dimension in the direction of loading.

With a little imagination one can see that buckling may be a relevant mode of failure in compressively stressed regions of thin plate and shell structures (*Figure 4.33*); for example one might mention compression flanges and web regions of I-beams, box structures or cylinders subject to bending or torsion, road tankers and vacuum tanks.

Here, for illustrative purposes only, an example of column buckling will be examined (see *Figure 4.34*). A column under compressive load may be short or long with respect to its cross-sectional size. Obviously

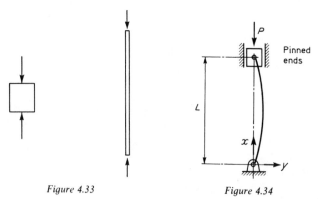

Figure 4.33   *Figure 4.34*

only the long case is of interest. In what follows, two assumptions will be made. First, the material is linear elastic and, second, when the column deforms it does so into a shape with 'small' curvature, i.e. only small displacement theory will be used. Of course with the short column no curvature would be expected at all but here the possibility of some curvature will be assumed at some load P. Then the bending moment *in the deflected form is* $M_x = Py$ where y is the displacement at any section distance x from one end. Notice that the above expression has been taken in the *deformed* condition. This may be looked upon as a first approximation to considering large displacement. Now, from the beam equation (4.7),

$$\frac{M}{EI} = \frac{1}{R}$$

where $1/R$ is the curvature of a beam. The expression for curvature in the x, y coordinate system shown is

$$\frac{1}{R} = -\frac{d^2y/dx^2}{[1+(dy/dx)^2]^{3/2}}$$

which for small curvatures gives in the above situation

$$EI \frac{d^2y}{dx^2} = -Py$$

or

$$\frac{d^2y}{dx^2} + n^2y = 0 \quad \text{where} \quad n^2 = P/(EI)$$

The solution to this linear differential equation may be found as

$$y = A \sin nx + B \cos nx$$

which can be checked by substitution. This expression leads to a nonlinear relation between load (P or n) and displacement (y) although a linear stress–strain relation has been used. The displacement boundary conditions for the pin-ended column shown are at $x = 0, y = 0$, and at $x = L, y = 0$, whereupon $A = 0$ and $B \sin nL = 0$, from which one may say either $B = 0$ or $\sin nL = 0$.

Notice that if $B = 0$ then $y = 0$ and there is no transverse displacement anywhere, indeed no buckling. However if $\sin nL = 0$ then B is undetermined and may have any value. Thus y is undetermined and can have any value when $\sin nL = 0$. This is the buckling condition and occurs when $nL = 0, \pi, 2\pi$, etc. Taking the smallest valid value (π) and denoting this load by P_c, the critical load gives

$$P_c = \frac{\pi^2 EI}{L^2}$$

which is frequently referred to as the Euler critical load after Euler (1707–1783). When $P < P_c$ then $\sin nL \neq 0$ and $B = y = 0$.

The rather strange load–displacement behaviour suggested by this solution is illustrated in the top curve of *Figure 4.35*. It should be noted that the buckling load is *unrelated to the static strength* of the material. However, the theoretical maximum load does depend on

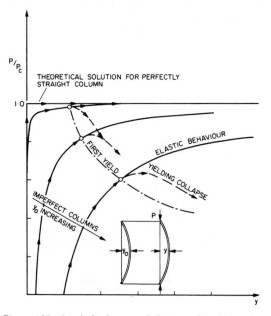

Figure 4.35   Load–displacement behaviour of buckling column

the degree of end fixing—clamping the ends against rotation increases the buckling load by a factor of 4 for example. Other buckling formulae for typical components and end-fixings are given in reference 4.

In practice, components are never perfectly flat or straight, especially if they have been fabricated by a welding procedure which introduces distortion. If an initial curvature before loading is assumed—expressed in terms of a central offset y_0—the lower curves in *Figure 4.35* are obtained, these being more typical of real buckling behaviour. Also, compressive yielding and collapse may intervene before the buckling load is reached. The load-carrying potential of the column is therefore much reduced by initial shape imperfections. The role of compressive residual welding stress has also been discussed in Section 1.4.

The behaviour illustrated by this example is typical of more complicated structures, and it should therefore be clear that in general where there is a risk of buckling, it may be no safeguard to increase the static strength of the material used. For a given cross-sectional area of component, it will instead be better to redistribute the material to produce a higher second moment of area, for example by making a cylindrical bar into a thin tube. However there are dangers even in this approach, as there is a limit to how thin part of a cross section may become before *local* buckling of the wall becomes important as distinct from overall buckling of the component. Local buckling cannot be dealt with here, but may be important in tubular structures, and in wide-flanged beams, etc.

4.1.10 More complicated stress analysis problems

The three simple basic principles of equilibrium, compatibility and stress–strain relations are still necessary and sufficient for the most complicated kind of static stress analysis problems that will be encountered but the mathematical complexities involved in treating even apparently simple shapes and loadings can become overwhelming. Nevertheless, many areas of practical interest relevant to welded structures have been explored by generations of mathematicians and engineers and it is important to know where such solutions are to be found, what their bases are and how they relate to the basic principles.

One important approach is to formulate a set of differential equations which simultaneously express the equilibrium and compatibility requirements, linked by an appropriate constitutive law (usually linear). Provided that there are not too many variables in the equations, these may be solved analytically in a form which allows the user to vary specific boundary conditions of loading or deformation. This basic technique tends to become specialised to particular component types—for example, shell theory (Novozhilov[24]), plate theory (Timoshenko and Woinowsky-Krieger[25]), buckling (Timoshenko and Gere[26]) and two-dimensional elasticity (Timoshenko and Goodier[27]) which deals with a variety of shapes. Such solutions are often termed 'exact'.

Algebraic equations are easier to solve than differential equations and *finite difference* is a technique which reduces the differential equations to algebraic equivalents and attempts to ensure that these are satisfied at a discrete number of points in a mesh covering the area of the component of interest.

Energy methods also provide a powerful alternative route to solution. Broadly there are two lines of attack:

1 The total potential of the loaded structure is first derived from a known or guessed compatible deformation/strain state using the appropriate constitutive relation. (The total potential is given by the strain energy less the potential of the external loading, this being the total change in energy state due to loading as the structure has gained some strain energy and the loads have lost some potential energy.) If this is then minimised with respect to appropriate variables, according to established theorems, a description of the true (equilibrium) state can be approached, and also a relationship found between the load and the displacement. Alternatively the loads may be found by equating external work done to internal strain energy stored.

2 The total complementary potential is derived from an equilibrium load/stress state and is then minimised to approach the true compatibility state.

Computer-based minimisation techniques have made energy methods very suitable for problems which could not previously be solved easily and use of both methods 1 and 2 above on the same problem (for single loads) can allow bounding of the answers. Variations on these techniques masquerade under different names (e.g. Castigliano's method, virtual displacement (or load), unit load, Rayleigh–Ritz or Galerkin methods) and it is often difficult to determine the optimum method for a given problem or indeed to ascertain what assumptions are being made in a given application. It is always worth questioning in a given instance whether equilibrium or compatibility is being satisfied within the material and/or at the boundaries of the structure.

A further application of energy-type principles is made in the *finite-element* method, whereby the structure is split up into a number (usually a large number) of discrete elements. The internal strains or stresses (sometimes both) are related to the displacements or forces at the *nodes* (or corners) of the elements and the strain energies or similar quantities calculated. The unknown nodal quantities are then found using an energy principle and this usually yields a large number of simultaneous equations which are normally solved using a matrix method, whereupon the resulting stresses and displacements can be evaluated on a point-to-point basis.

Unfortunately, the complications of the techniques make it difficult to ensure that compatibility is properly satisfied, i.e. there may be 'chinks' of light between the elements or slope discontinuities between them. On the other hand equilibrium may not be properly satisfied where the corresponding complementary energy route is used. It is therefore often impossible to show that the answers are true bounds. However, the method is now widely used in the design of components and structures, mostly through the application of computer packages bearing such acronyms as PAFEC, BERSAFE, NASTRAN, MARC, ANSYS, STRUDL (see Hutton and Rostron[28]). These packages spare users the complexities of mathematical manipulation and offer a range of procedures to facilitate painless and minimum-error input and output of data. Usually different types of of element suitable for different kinds of problem are offered. For example, a pressure vessel stress analysis may be carried out using 'shell' elements whereas a space-frame structure analysis could be carried out using simple 'beam' elements.

As with most analytical procedures which require the user to hand over control of the computational process, considerable care and experience is necessary for proper application of the finite-element method in the general case and most experienced users advocate careful cross-checking of the results by comparing with simple analyses which can be carried out 'by hand'. Particular care needs to be taken with element size which needs to be refined in areas of high strain gradient and with the specification of boundary constraints.

4.1.11 Limit of elastic behaviour and yield criteria

If linear elastic stress–strain laws are used, one must also know the maximum stresses and strains for which they are valid. Usually the yield stress and strain of a material is obtained during the course of routine tensile testing, but this value will only be valid for the test conditions, i.e. valid for uniaxial normal tensile stress. The problem is left of predicting yield conditions for shear stresses and other multiaxial stress systems.

Tresca recognised in 1868 that yielding was a shear-operated process, and therefore proposed that when the maximum shear on any plane in a multiaxial stress system equalled the maximum shear experienced during a simple tensile or compression test, yield would occur. As the maximum shear stress is given by the maximum principal stress difference, (from the Mohr circle, for instance) Tresca's criterion says that yield occurs when

$$|\sigma_1 - \sigma_2| \quad \text{or} \quad |\sigma_2 - \sigma_3| \quad \text{or} \quad |\sigma_2 - \sigma_1| \geqslant \sigma_Y$$

The Von Mises–Hencky–Huber–Maxwell criterion proposes that shear stresses on planes other than the maximum exert an influence, and that yield occurs when

$$\frac{(\sigma_1 - \sigma_2)^2}{2} + \frac{(\sigma_2 - \sigma_3)^2}{2} + \frac{(\sigma_3 - \sigma_1)^2}{2} \geqslant \text{constant}$$

The constant may be found by substituting the principal stresses at yielding under simple tension into the function, whereupon

$$\text{constant} = \frac{(\sigma_Y - 0)^2}{2} + 0 + \frac{(0 - \sigma_Y)^2}{2}$$

and the Von Mises criterion becomes

$$\frac{1}{\sqrt{2}} [(\sigma_1 - \sigma_2)^2 + (\sigma_2 - \sigma_3)^2 + (\sigma_3 - \sigma_1)^2]^{1/2} \geqslant \sigma_Y$$

or, in terms of the stresses on a 90° element,

$$\frac{1}{\sqrt{2}} [(\sigma_x - \sigma_y)^2 + (\sigma_y - \sigma_z)^2 + (\sigma_z - \sigma_x)^2 + 3(\tau_{xy}^2 + \tau_{yz}^2 + \tau_{zx}^2)]^{1/2} \geqslant \sigma_Y$$

$$(4.26)$$

In the common situation where the stress system is biaxial ($\sigma_3 = 0$) the Tresca and Von Mises criteria may be represented graphically in nondimensional form as in *Figure 4.36*. This figure is usually referred to as the yield surface plotted in stress space.

It can be shown that the Von Mises criterion is proportional to the strain energy of distortion and is thus sometimes thought of as an energy criterion being fulfilled. However, as yielding is so closely connected with shear, it is perhaps more meaningful to think of it as related to the root-mean-square of the shear stresses. Its use is justified principally because it seems to agree with experimental findings.

The simplest specimen for experimental determination of the yield surface is the thin, constant thickness, circular tube. Various types of loading lines, e.g. axial load, torque, pressure and combinations of pressure and axial load, are shown in *Figure 4.36*. Note that hydrostatic pressure ($\sigma_1 = \sigma_2 = \sigma_3$) has no effect on either criterion and it is generally accepted that it has little effect in practice. If $\sigma_1 = \sigma_2 = \sigma_3$ yielding does *not* occur. Assuming this to be the case and assuming that compressive yield is identical to tensile yield, it is possible to establish the yield surface with very little testing.

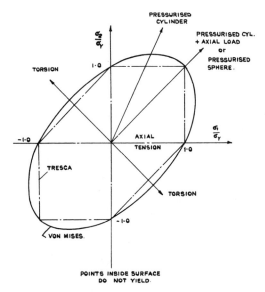

Figure 4.36   Yield surface plotted in stress space

4.1.12 Experimental methods

The foregoing sections have been almost exclusively concerned with *theoretical* methods of stress analysis. A large number of techniques are available for the *experimental* determination of strains and stresses on full-size components or models where calculations are impossible or need substantiation.

Many books have been written on this topic alone and the reader is referred to one of the many handbooks on experimental stress analysis. Some of the more common techniques are extensometry, electrical resistance strain gauging, strain measurements via surface coatings, photoelasticity, holography, etc. Most of these and others can be employed in static or dynamic situations and even on occasions at moderate temperature levels. All can be accurate but care and experience are essential.

It is perhaps worth mentioning that photoelasticity is the only experimental method which may be used to measure *stresses* directly and furthermore can indicate stresses within a component. Most other methods measure strains—usually on the surface—and stresses are inferred via the relevant stress–strain relationships. The effect of fillet and overfill shape on weld stresses has been investigated on photoelastic models of welds[22].

A word of caution may be relevant against thoughtlessly embarking on experimentation. The design of an experiment and the modelling of a problem for experiment is something that demands careful consideration. Many experiments in the past have been no more than an analogue of the mathematics. As such they are largely worthless as a means of checking the assumptions made in a theory. See for example reference 7. Broadly speaking, experiments should be aimed at elucidating the assumptions made theoretically or at problems where theory is not available or not directly applicable.

Scale models are often used in engineering experiments and can provide useful final checks on a proposed design. However, one must remember to scale loads, pressures, etc., in accordance with the principles of dimensional analysis (see reference 23). Welded models have their own problems due to the practical difficulties of producing tiny welds. Features such as weld shape, residual stress and cracks, cannot usually be incorporated in the model according to the previously mentioned dimensional principles.

4.2 Sizing welds

At some stage in the design procedure, decisions must be taken on the size of welds. It is probably obvious that stress calculations will be used as a basis for these decisions with the proviso that the resulting welds can be produced soundly in practice, and that they do not appear to be ridiculously large or small relative to the rest of the component. However it should be clear from the preceding section that a precise calculation of stress in welds will hardly ever be possible, as the principles of stress analysis can be difficult to work out with any rigour, even in a component of simple geometrical shape.

The purpose of this section, therefore, is to make rough estimates only of the nominal stress levels in welded joints by comparing their shapes and configurations with the simple cases condensed into *Figure 4.18*. Stress concentration effects due to notches (see Section 4.3) are temporarily ignored, and the weld size is adjusted until a so-called 'permissible' or 'design' stress is not exceeded. Direct comparison with material strength properties such as yield or ultimate strength would not be realistic as the calculation will be very approximate.

In view of these remarks one may be inclined to ask whether such rough calculations are worth doing at all. Alternatively, what is the basis of the 'permissible' stress, if it is not a measurable material property? Probably, the main justification for the nominal stress calculation is that it introduces a measure of uniformity into weld sizing procedures, and allows one to make use of a considerable body of previous (one hopes satisfactory) experience which is based on a similar approach.

Permissible stress

A recommended permissible stress level, whether it is related to a material property or not, characterises successful design practice in a particular class of structure and loading. However, the obvious trap which everyone falls into at some time or another, is to transfer or extrapolate a recommended value from one class of structure to another class where the loading, the environment, the workmanship or the material response may be different. The most obvious illustration of such a mistake is the application of design stresses suitable for steady loading, to structures where the load fluctuates and fatigue failure becomes possible.

There is no reason, therefore, why different codes of practice should agree on design stresses for welds, nor should we expect an obvious rational basis. However, there are some common features. Most code design stresses are expressed as a proportion of the material yield or ultimate strength, except where fatigue loading is possible (see Chapter 6). The danger of the yield strength basis is that it encourages the application of materials with a high ratio of yield/ultimate strength, which may therefore be used with little reserve of ductility.

The design/yield stress ratio is often varied depending on the character of the stress. A statically determinate direct or membrane stress which could lead directly to collapse over large and important regions of the structure is usually held to a lower level than a bending or shear stress for which local yielding would be of lesser consequence.

Normal and shear stresses acting at the same point are usually combined in terms of 'equivalent' or Von Mises stress. Paradoxically, shear stress acting alone is sometimes allowed to rise to a higher equivalent stress. For example, BS 153[8] allows bending stress or combined equivalent stress to reach 66% of yield, whereas equivalent stress for shear acting alone is tolerated to 76%. Direct or 'primary' stresses on the other hand are held to 59% of yield.

If the weld is located in a region of compressive stress in a thin component, the question of buckling arises, and the permissible stress may then be governed as a proportion of the theoretical buckling strength.

Other points concerning permissible stress will be discussed in the following section as appropriate.

Analysis of butt welds

A butt weld can be considered to be an integral part of the loaded component, especially when full penetration is specified. There is, therefore, no special analysis problem—if the component can be analysed, then the weld stresses are provided as a consequence. It is sensible to ignore joint overfill (traditionally and misleadingly called 'reinforcement') because it does not contribute to the joint strength; in fact it has the reverse effect by providing a stress concentration. For partial penetration welds, the calculation should be based on the welded ligament only. In many standards, even where fatigue is unlikely, partial penetration welds are discouraged for primary tensile loads, presumably because they are prone to cracking in manufacture, or in other ways increase the risk of fracture in service.

In most materials, it is possible to match the strength of the weld metal and associated zones to the parent material and for this reason the permissible stresses for butt welds are usually the same as specified for parent materials. However, in some materials, difficulties arise because the parent material strength depends on work hardening, quenching and tempering, or ageing, and a subsequent fusion weld either fails to match this strength or produces softened zones. This only need be an embarrassment in 'primary' load locations, for example at transverse butt welds where yielding would lead to collapse. In other places, for example at stress concentrations, weld ductility and toughness are probably more valuable than static strength.

Analysis of fillet welds

By contrast, fillet welds are nonintegral in character, and have a shape and orientation relative to the loading which almost disallows the application of simple stress analyses.

Two simplifying assumptions are implied in most practical design methods for fillet welds; the first is that the stresses ought to be related to the weld cross-sectional area. Even the cross-sectional area is nonuniform, however. Hence the minimum cross section or 'throat' plane is chosen as a representative or reference area. For 45° fillet welds the throat area is given by 0.707 × leg length × weld length.

Secondly it is assumed that the throat plane does not lie at an angle to the load or to the plane of the component, but is parallel or perpendicular to it. Thus the fillet welded joint is rearranged to resemble a butt weld in which the ligament area is given by the throat area (see *Figure 4.37*).

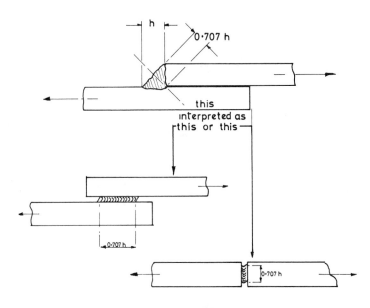

Figure 4.37

The inaccuracies of this approach are recognised by reducing the permissible stress. Values recommended in BS 153, and in other codes of practice relating to structural steel, set the allowable stress on a fillet weld to a value of about $100 \, \text{MN/m}^2$ irrespective of yield strength. This may be reduced further for load-carrying fillet welds subject to fatigue. The distinction between shear and normal stress is irrelevant because of the assumption in the previous paragraph. It is interesting to note that $100 \, \text{MN/m}^2$ corresponds to the allowable steady shear stress in parent material or butt welds made of minimum yield strength mild steel.

These principles may now be illustrated in a series of examples:

Example 4.4: Load parallel to fillet welds (Figure 4.38a)

Neglecting the small bending moment acting at the weld plane, and assuming that the shear stress is uniform along the length of the weld,

$$\tau_{average} = \frac{P}{2 \times 0.707h \times l}$$

The assumption of uniform stress requires some discussion. If the plates are thick relative to the weld, they may be assumed to be rigid,

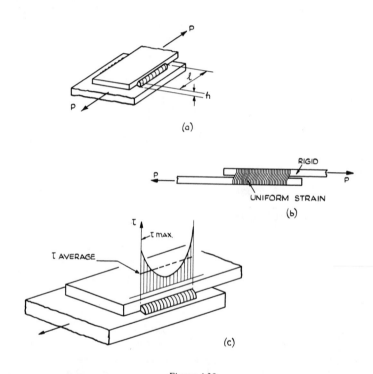

(a)

(b)

(c)

Figure 4.38

and compatibility requires that the strain and consequently the stress is uniform along the weld length, as shown in *Figure 4.38b*. If, on the other hand, the plates strain appreciably, a greater proportion of the load will be transferred near the weld ends where the strain is high, giving rise to the shear stress distribution shown schematically in *Figure 4.38c*. It is possible to quantify this effect, as in reference 9. The analysis given there is useful for the examination of stress trends, although it is a somewhat sophisticated treatment of what is after all a rather crude joint. The maximum shear stress at the weld ends is given by:

$$\tau_{max} = \tau_{average} + \frac{\sigma \lambda l}{12 \times 0.707h}$$

σ is the nominal stress in the plate and λ is the stiffness of a unit length of weld (determined experimentally) and therefore depends on the throat dimensions.

If σ is low (thick plates), the second term above is small and the assumption of uniformly distributed shear stress is reasonable. Although $\tau_{average}$ can be reduced by increasing the weld length the influence of 'l' in the second term leads to a 'peakier' stress distribution. However, a later figure in the same reference demonstrates that short side fillet welds will also have disadvantages in that the normal stress *in the plates* will be elevated near the edges.

It is unfortunate that the stress peaks should arise at the weld ends, because this region will usually contain other small-scale stress raisers due to the shape of the weld end, and the presence of stop/start defects. For this reason, it is considered good practice to continue the weld for a short distance around the corner, and along the end faces of the plates (see also *Figure 6.8*).

Example 4.5: Lap joint—end fillets (Figure 4.39)

Again ignoring bending, and assuming the stress distribution to be uniform, the average shear stress is given by

$$\tau = \frac{P}{2 \times 0.707hl}$$

If the end fillet covers the full width of both plates, there is no reason for the stress to vary across the width. If the weld is shorter, the

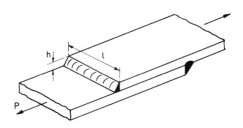

Figure 4.39

stress pattern will probably be similar to that of a stepped width plate[10], that is, uniform in the central region with stress peaks at the ends.

Example 4.6: Weld all round (Figure 4.40)

The average shear stress is again given by

$$\tau = \frac{P}{\sum 0.707hl}$$

It is clear from the results of fatigue tests that stress peaks still occur at the plate corners. Rounded corners will, of course, help the welder but do not seem to reduce the stress peaks greatly. The author has

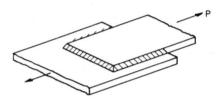

P

Figure 4.40

no positive information on the proportion of load transmitted through the end fillets versus the side fillets, but one imagines that heavy end welds might help to reduce the stress peaks from Example 4.4. It is also good practice to weld the end fillets first while the lap plate is less restrained.

Example 4.7: Bending and shear of fillet welds (Figure 4.41)

If a significant bending moment is developed at the weld plane, different in- and out-of-plane shears will be developed (on the assumption that the throat plane is perpendicular to the weld plane).

The maximum out-of-plane shear due to bending appears at the weld ends and is given by equation (4.7) as

$$\tau_{bending} = Pd \times \frac{y}{I} \quad \text{where } I \text{ is the second moment of area of the weld group}$$

$$= Pd\frac{12}{2 \times 0.707hl^3}\frac{l}{2} \tag{4.27}$$

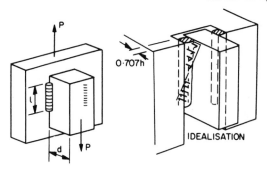

Figure 4.41

(Note that equation (4.27) will also give the same value for *normal* stress due to bending, if the throat plane is assumed to be rotated 45° *into* the plane of the joint.)

The in-plane shear stress distribution on the other hand is altered by bending to the extent that it reaches a maximum at the centre and drops to zero at the weld ends. Hence from equation (4.20)

$$\tau_{max} = \frac{PA\bar{y}_{max}}{bI}$$

$$= \frac{P \times 2 \times 0.707h \times (l/2) \times l/4}{2 \times 0.707h \times (2/12) \times 0.707hl^3}$$

$$= \frac{3P}{4 \times 0.707hl} \tag{4.28}$$

This treatment assumes that the components extend sufficiently far out of the weld plane, that the complementary shears implied by equation (4.28) can be sustained in this direction.

The in- and out-of-plane shear stresses at any distance *y* from the neutral axis are obviously assumed to act on the same cross-sectional area and can therefore be added vectorially, i.e.

$$\tau_{total} = (\tau_b^2 + \tau_s^2)^{\frac{1}{2}}$$

Note that if the bending stress is treated as a normal stress, then the addition of stresses would strictly have to be made by Mohr's circle as in the example on page 164. However, the numerical difference in the result is insignificant when considered relative to the gross inaccuracies in other respects.

Example 4.8: Attachment of unsymmetrical sections (Figure 4.42)

Space framed structures (see *Figure 4.1*) are often analysed as if the connections between the members were pin jointed. This assumption reduces the structure to a statically determinate case which can be analysed largely by equilibrium. If welded connections are used, the attachment should be arranged so that the joint does not rotate, nor does the attached member bend when it is loaded axially, thus preserving the fictional statically determinate analysis.

The requirements therefore are:

1 The displacement (not the strain) of each weld attachment point should be the same.
2 The load P is applied through the centroid of the section.

From equilibrium

$$P = P_A + P_B$$

leading to

$$P_A = \frac{Pb}{a+b}, \qquad P_B = \frac{Pa}{a+b}$$

$$P_A a = P_B b$$

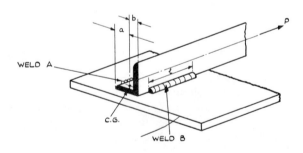

Figure 4.42

Therefore the shear stress on weld A is given by

$$\tau_A = \frac{Pb}{a+b} \times \frac{1}{0.707 h_A l_A}$$

and the shear stress on weld B is given

$$\tau_B = \frac{Pa}{a+b} \times \frac{1}{0.707h_B l_B}$$

Thus the shear displacements are

$$\delta_A = \frac{Pb}{a+b} \times \frac{1}{0.707h_A l_A} \times \frac{1}{G} \times 0.707h_A$$

and

$$\delta_B = \frac{Pa}{a+b} \times \frac{1}{0.707h_B l_B} \times \frac{1}{G} \times 0.707h_B$$

and from condition 1 the weld lengths should be in the ratio

$$\frac{a}{b} = \frac{l_B}{l_A}$$

4.2.1 Welds in torsion

Two quite separate idealisations may be made when considering groups of welds subject to torsion. We may assume (i) that a weld will support shear stresses due to torsion in any direction, or (ii) that it can only support shear along the weld line. In deciding which assumption is valid, it is useful to remember that shear cannot be supported normal to a free surface. Assumption (i) is tantamount to saying that the components being joined are relatively rigid. The distinction between the two situations is shown in *Figure 4.43*. The two idealisations would lead to the same conclusion in the case of a circular attachment weld (*Figure 4.44*).

Example 4.9: (Figure 4.45)

According to treatment (i), the total force on each element of weld consists of a vertical force giving a vertical shear stress

$$\tau_{DIRECT} = \frac{P}{\sum 0.707hl}$$

and a torsion shear stress.

The torsion stress at each point is dealt with in the following approximate fashion. The force on each element is assumed to be

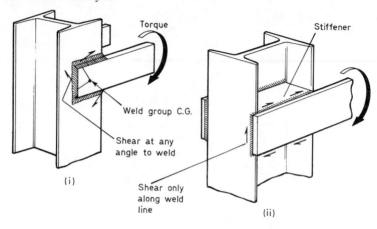

Figure 4.43

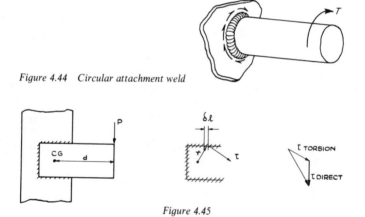

Figure 4.44   *Circular attachment weld*

Figure 4.45

proportional to the distance from the centroid of the weld group. (This is reasonable, as the strain is probably proportional to the distance r. Hence $r \propto r$.) Torque resistance of a small element, δl

$$= \tau r \, dA$$

$$= \frac{\tau}{r} r^2 \, dA$$

$$= \frac{\tau_{max}}{r_{max}} r^2 \, dA$$

Therefore total torque resistance of weld group

$$= \frac{\tau_{max}}{r_{max}} \times \int_{weld\ length} r^2 \, dA$$

$$= \frac{\tau_{max}}{r_{max}} \times J$$

(where J is the polar second moment of area)

(*Note:* J = Area $\times \sum [r^2 + (l^2/12)]$ by parallel and perpendicular axis theorems.)
Hence

$$\tau_{max} = \frac{P \, dr_{max}}{\sum 0.707 lh[r^2 + (l^2/12)]}$$

and acts perpendicular to the radius r.

The stress due to torsion is combined vectorially with that due to direct shear and the result compared with allowable stresses.

Example 4.10

According to assumption (ii), the torque is calculated by multiplying the vertical force P by the distance to the shear centre. (See *Figure 4.18*.) The maximum shear stress due to torque is found (*Figure 4.46*) from

$$\tau_{max} = \frac{3Pe \times 0.707 h_{max}}{\sum [l(0.707h)^3]}$$

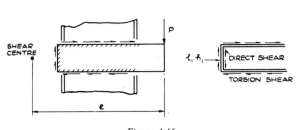

Figure 4.46

Direct shear can only be developed in the vertical weld group and is given by

$$\tau_{DIRECT} = \frac{P}{0.707 h_1 l_1}$$

This is acting along the same line as the torsion stress and can, therefore, be subtracted directly from it.

Finally, a word of warning on the problem of softened zones adjacent to fillet welds.

In certain materials for example Al–5%Mg alloy, soft head-affected zones are produced. In the case of fillet welds, transverse to a primary load, the softened zones may extend through the material thickness and allow failure of the joint, independent of the fillet weld size. An interesting investigation of fillet weld details in such materials is reported in reference 11. An alloy which age-hardens 'naturally' (that is at service temperatures), may allow the full parent material strength to be developed in the softened zones some time after welding (see also page 117).

Arrangement of spot welds (*Figure 4.47*)

The size and disposition of spot welds in a given joint are determined as much by the requirements of the welding process, as by consideration of joint strength. The weld diameter d is usually related to the material thickness t by the equation $d = \sqrt{t}$. Spots should be spaced

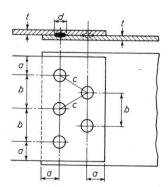

Figure 4.47

above a certain minimum to avoid electrical shunting of current, and welds should be placed sufficiently far away from plate edges to reduce the risk of edge tearing. The main recommendations in these respects are summarised in *Table 4.1*, due to Koenigsberger[12].

Table 4.1

						b max	
t		d min	a min	b min	c min	Single row	2 or more rows
(swg)	(in)	(in)	(in)	(in)	(in)	(in)	(in)
24	0.022	0.125	0.188	0.5	0.5	0.5	
22	0.028	0.188	0.313	0.563	0.563	0.563	0.625
20	0.036	0.188	0.313	0.563	0.563	0.563	0.625
18	0.048	0.188	0.313	0.563	0.563	0.563	0.625
16	0.064	0.25	0.375	0.75	0.75	0.75	1.125
14	0.080	0.313	0.5	0.937	0.937	1.25	1.875
12	0.104	0.313	0.5	0.937	0.937	1.25	1.875
10	0.128	0.375	0.625	1.125	1.125	1.5	2.25
	0.188	0.437	0.687	1.313	1.313	2.25	3.375
	0.25	0.5	0.75	1.5	1.5	3.0	4.5

The strength of joints is usually difficult to define for a number of reasons:

1 The load is transmitted discontinuously through a given pattern of spots, those in high strain areas transmitting more than others
2 The effect of plate friction is unknown
3 A sharp crack effectively exists at each spot
4 The mechanical properties of the material surrounding each spot cannot be easily estimated

Problems 1 and 2 are common to rivet joints which have been designed relatively successfully in the past, on the assumption that the load is shared equally by all rivets, and that it will be conservative to ignore plate friction. The same philosophy is carried over to spot welds which are sized on the assumption that the carrying capacity of each spot should be related to its cross-sectional area.

It is generally observed that cracks propagate less readily in shear than in the 'opening' mode (see Chapter 5). Hence the joint is usually designed to avoid normal or 'peeling' loads, for which the strength is much lower.

Very few explicit recommendations on permissible stress levels are known to the author. One aircraft manufacturer, at least, recommends avoidance of spot-welded joints for primary loadings and advises design stresses in the range 45 to 75 MN/m^2, the lower end of the range being associated with large thickness or joints between plates of unequal thickness. One suspects that in a given case, joint failure depends so much on items 1 and 4 above, that the designer really needs first-hand information from tests on full patterns of spot welds which are carried out by the fabricator in question.

The following examples serve to illustrate the calculation of nominal stresses by methods which are similar to those used for riveted joints.

Example 4.11   Spot welded joint (direct force)

Making the assumption (a) that the load is shared equally by all welds, and (b) that the bending moment may be neglected,

$$\tau = \frac{P}{n \times \pi d^2/4}$$

where n is the number of spots.

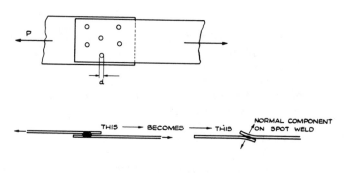

Figure 4.48

One suspects that the end rows of spot welds will carry more of the load for the same basic reasons as given in Example 4.4.

It can be shown (Koenigsberger[12]) that the bending moment which could be induced in plates less than 1.5 mm thick would raise the bending stress in the plates to yield point (at least in mild steel), allowing the behaviour illustrated in *Figure 4.48*. Significant normal stresses would then be developed. This situation may be avoided by having more than one row of welds.

Example 4.12   Spot welded joint (moment) (Figure 4.49)

The reasoning in Example 4.9 may be employed here. We assume that a shear stress may be developed by a spot in any direction and

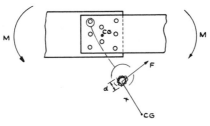

Figure 4.49

that the shear stress developed by a particular spot will be proportional to its distance from the centroid of the weld group. Hence

$$\frac{\tau}{r} = \text{constant} = C, \text{ say, and } \tau = \frac{F}{\pi d^2/4} = Cr$$

where τ is the shear stress on any spot, F is the shear force in it and r is its distance from the centroid. The torque developed by a spot will be given by

$$\Delta T = Fr = \tau \times \frac{\pi d^2}{4} r$$

$$= C \frac{\pi d^2}{4} r^2$$

and the total torque developed will be

$$T = C \sum \frac{\pi d^2}{4} r^2$$

Hence the shear stress at each spot

$$\tau = \frac{T}{\sum \frac{\pi d^2}{4} r^2} r$$

The relevance of rupture test results

For fillet and spot-welded structures especially, it may be comforting to have information derived from full-scale loading tests which are taken to fracture. In a well-designed joint which has been welded without disadvantage to the ductility of the various zones, it is common to find that failure occurs in the parent material at nominal stresses corresponding to a high proportion of the ultimate strength. This is encouraging, but *strictly only relevant to structures which are loaded to the point of collapse in service.* (One imagines that such structures are rare.) In fact the preceding nominal stress calculations which assume uniform stress in the joint are probably nearer the truth where rupture has been preceded by plastic flow. In such cases load is shed from the highly stressed regions. For structures which are loaded in service to a fraction of the collapse load (say 25%), rupture tests merely draw attention to gross errors in sizing, or areas where the ductility or the ultimate strength has suffered during welding. These tests have no relevance in the case of a fatigue loaded structure.

Welds at discontinuities

For obvious reasons we often wish to have a weld at a change in component geometry (see *Figure 4.50*). A general approach to such cases may be outlined:

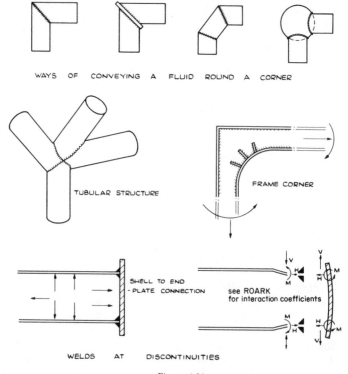

Figure 4.50

1 Estimate loads and moments across the joint, assuming complete strain and displacement compatibility between the components to be joined.
2 Apply these moments and loads to the proposed weld shape in the manner described in the previous examples.

As an illustration of this general approach, we might examine a nozzle/sphere pressure vessel joint in which partial penetration welds are used (see *Figure 4.51*). This design detail has been the subject of controversy in the power station boiler field.

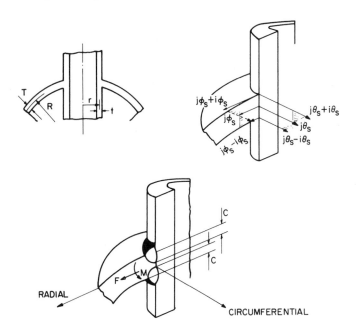

Figure 4.51

Example 4.13   Nozzle attachment weld

A cylindrical nozzle of 28 in bore and 2 in thick is fitted to a spherical shell of 145 in inside diameter and 5 in thick. Find the stresses for various sizes of partial penetration welds.

$R = 75\,\text{in}$, $r = 15\,\text{in}$, $r/R = 0.2$, $p = 3000\,\text{lbf/in}^2$
$T = 5\,\text{in}$, $t = 2\,\text{in}$, $R/T = 15$

From Rose and Thompson[13] the stresses in the radial direction at the nozzle are given in terms of coefficients $j_{\phi s}$ and $i_{\phi s}$.

$$\sigma_{r\ direct} = j_{\phi s}\,\frac{pR}{2T}$$

$$\sigma_{r\ bending} = i_{\phi s}\,\frac{pR}{2T}$$

Therefore the radial force/unit circumference F is $j_{\phi s}(pR/2T)T$ and the radial moment/unit circumference M is $i_{\phi s}(pR/2T)\frac{1}{6}T^2$. Hence the stress in the partial penetration welds neglecting the fillet is

$$\left(\frac{F}{2C}\pm\frac{My}{I_{weld}}\right)$$

Therefore

$$\sigma_{total} = \frac{pR}{2T}\left(\frac{j_{\phi s}T}{2c} + i_{\phi s}\frac{T^2}{6}\times\frac{T}{2\times 2\left[\frac{1}{12}C^3 + C\left(\frac{T-C}{2}\right)^2\right]}\right)$$

From Figure 14 of reference 13 $j_{\phi s} = 0.4$; $i_{\phi s} = 0.1$ for the given ratios r/R and R/T.
   Hence:

Weld penetration c	σ_{total} (tonf/in^2)	Comments
2.5	5.0	complete penetration design
2.0	6.0	recommended in BS 1515
1.5	7.7	
1.0	11.3	

$\Big($ Note also that the maximum stress in this nozzle is circumferential and is given by

$$(j_{0s} + i_{0s})\frac{pR}{2T} = 26\,\text{tonf/in}^2\Big)$$

This calculation might be repeated to include the fillet throat.
   The stresses associated with the BS 1515 design seem reasonable in terms of permissible stresses, but the relevant question in the assessment of the design is 'Will the crack in the unwelded zone propagate?' (See Chapter 5.) Of course the completely penetrated weld design causes the least worry from the service performance aspect. However, the weld edge preparation needs to be wide enough to allow access to the root and this will no doubt lead to massive fillets and large volumes of deposited metal. As discussed in Chapters

1 and 2, this will increase the tendency to distortion, high residual stress and fabrication cracking problems. Once again the designer is faced with an unhappy compromise.

4.2.2 Load flow concepts

In a number of situations, a rather intuitive approach is often used, involving the idea of loads 'flowing' across a welded connection. It may be helpful to investigate the basis for this approach, because it could lead to false conclusions in the hands of a designer lacking natural intuition. A typical situation is considered below, and comments are made (in brackets) from the fundamental point of view on the various assumptions which are commonly made by designers.

Example 4.14

A frame consists of a number of horizontal I-beams connected in a fashion to be determined, to vertical I-beams (see *Figure 4.52a*). The loads are known.

The designer's first step would be to analyse the structure as a whole, assuming rigid connections between the verticals and horizontals, and thus obtain the moments, and horizontal and vertical forces adjacent to each connection.

(The designer has now committed himself to a joint which ensures complete compatibility of rotation and displacement between the end of the horizontal member and the cross section of the vertical member. See step 2 of 'Welds at Discontinuities' on page 197. In other words, he has assumed that the weld is rigid.)

To simplify calculations he might replace the moment M by flange loads F, where $M = 2Fh$.

(The assumption that the web plays a negligible part in transmitting a moment is justified in the case of typical I beams in which the second moment of area of the web is negligible compared with the flanges; this should be checked!)

He would add flange loads P due to the horizontal load H, assuming that this load is uniformly distributed over the cross section of the horizontal I beam.

(The load is distributed uniformly only if the end of the horizontal beam is forced to *strain* uniformly over its cross section. (Compatibility.))

The trial design shown in *Figure 4.53b* might be considered. The welds A would be designed to transmit the total flange load at a 'permissible' stress level. The welds B would be designed to transmit

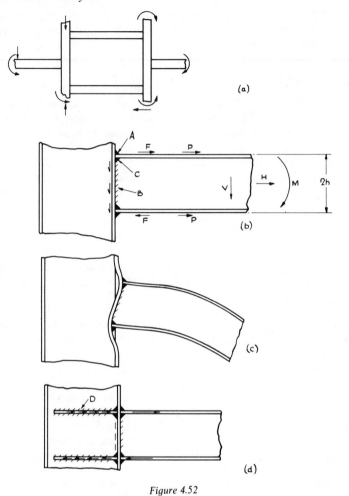

Figure 4.52

the vertical shear load (giving maximum shear stress at the neutral axis, see Example 4.7), and the combined shear stress level at the point C would be checked.

(The tendency in such a connection would be for the flanges of the vertical beam to distort as shown in *Figure 4.52c*. This simple fact would impinge directly on the assumptions made as follows:

1 The resulting joint flexibility would upset the original structural calculations leading to different loads and moments, which would be extremely difficult to estimate.

2 Most of the horizontal load *H* and moment *M*, would be transmitted through the web welds B, and not through the flange welds.

Turning back to the intuitive approach, the loads are required to 'flow' from the horizontal beam flanges into the web and diffuse through the web of the vertical beam into the flanges of the vertical.

All of this load flow takes place abruptly through a small cross section, and by analogy with fluid flow this design may be regarded as unhealthy.)

The designer whose intuition is well trained would at this stage add stiffeners (*Figure 4.52d*). This would restore rigidity to the corner and give the loads some room to diffuse. The welds D would be designed to transmit the flange loads to the web of the vertical. As rigidity has been restored to the corner, the original structural calculations would be valid. More information on the detail design of frame corners is given in reference 14.

4.2.3 Summary

In this section, the concept of a nominal, permissible, allowable, or design stress has been used in the manner of a guide for the designer who wishes to make a first estimate of weld size and disposition. This basis may seem to be more rational than an approach which relies on intuition, experience and the eye of the designer. It is not necessarily better, however. The user should strongly resist the inclination to assume that such a basis in conjunction with a large so-called 'safety factor' will automatically prevent failure of the structure. Real safety is only obtained when the detailed failure mechanism is known with a much greater precision than has been implied so far.

In Section 4.3 the effect of stress concentrations, residual stresses and yielding are discussed.

4.3 Stress concentrations, residual stress and yielding

The preceding sections have indicated how stresses may be calculated and have given some indication that stress fields in general are not uniform. Certainly at welds, other than flush butt welds, the stress pattern is rarely uniform and it is appropriate to describe such a region as a stress concentration.

4.3.1 Stress concentrations

The simplest example of a stress concentration is a tensile member with a sudden change of section (*Figure 4.53*). One can easily visualise the load flowing along the bar in a uniform pattern and then abruptly

Figure 4.53

bunching together to pass the narrow section. Wherever changes of section, notches, etc., occur which disturb the smooth flow of load (stress) the stress will be raised at least locally. Notice that a sudden increase in section can be almost as bad as a sudden decrease. It is usual in such cases to define a stress concentration factor (SCF) as the ratio of the maximum stress existing locally to the nominal stress existing in an undisturbed part of the section.

SCFs for a large number of components are available in the stress analysis literature. References 10 and 15 contain a large number of factors for simple beams, bars and plates with stepped sections, holes or notches. For example, for the case of an infinite plate in uniaxial tension disturbed by a small hole, the SCF is 3.0. Usually the factors are based on linear stress/strain behaviour.

Since SCF values can frequently be high (values from 1 to 10 are not uncommon) it is pertinent to ask what happens if the stresses in the highly stressed region cause yielding, and also if the calculated SCFs have any significance since they are based on linear elastic behaviour. Fortunately the answer to the second point is in the affirmative and it has been found that elastic calculations are useful not only in elastic behaviour, but also in limited plasticity, fracture

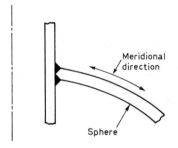

Meridional direction

Sphere

Figure 4.54

via fracture mechanics, and to some extent in creep. Linear stress analysis is thus a powerful tool not only for analysing stress but in studying the complete behaviour of engineering structures. The business of yielding will be considered in some detail.

In discussing stress concentrations it is useful to mentally (at least) distinguish between local and general SCFs. For example, at the toe of a fillet weld there might exist a rather high SCF which will likely be local in character having high stress gradients where the stress reduces over a short distance to the nominal level. Conversely at the junction of two parts of a pressure vessel such as a cylindrical shell and a head, or perhaps within a head such as a torispherical one, there may exist a moderate 'discontinuity' SCF which extends over the complete circumference. Much depends on the loading and service conditions as to whether these will be treated similarly or differently in design. One thing at least can be noted at this stage. Stress concentrations vary with the direction and type of loading. For example, the weld shape detail at a nozzle/sphere junction will be a significant *additional* stress concentration as far as the meridional or radial stress is concerned (*Figure 4.54*) but largely irrelevant for stress in the direction circumferential to the nozzle (reference should be made to the classification of weld details in Section 6.5.2). The word 'additional' was used deliberately because, of course, the stresses at the junction are already altered by the nozzle/sphere discontinuity from their nominal values remote from the junction. Hence the peak stress will be found by multiplying the two SCFs and the nominal stress in the unpenetrated sphere.

4.3.2 Yielding and beyond

Certain schools of thought in the past have maintained that SCFs at highly stressed regions were eliminated by yielding if the load was raised sufficiently and consequently they tended to put little weight on calculated elastic SCFs. This heresy led to the tendency to test components by loading them until failure occurred and then to say that the failure load divided by the design load was the safety factor. Such contentions are only true in the special case where the geometry of the component changes enough to alter its shape and consequently the stress distribution. (Technically, as described on page 163, significant alteration in shape occurs when the new shape is sufficiently different from the original to alter the equilibrium conditions. This is sometimes termed 'large displacement' or, in the case of cracks, 'crack blunting'.)

Similarly, the so-called safety factor on this basis is not a safety factor against any mode of failure except steadily increasing load applied once. Neither of these last two situations have much application in normal engineering structures. However, there is a grain of truth in the idea since, for an elastic/perfectly plastic material (such as mild steel), (see *Figure 4.4*), the *maximum stress* in a component can never become greater than the tensile yield stress (or its equivalent on, say, the basis of the Von Mises criterion) and the *minimum stress* can never be less than the yield stress in compression. The SCF remains the same unless the geometry alters sufficiently or the peak stress range as calculated from the SCF exceeds twice the yield stress. In the latter case the SCF would be better thought of as a strain concentration factor.

Shakedown

In modern design, local yielding at stress concentrations is usually considered permissible provided the plastic zone is fairly localised. If yielding were to be prohibited, most components would be rather larger or stress concentrations would have to be reduced to a very low order (a pious hope as far as welded fabrications are concerned).

Effects that need to be borne in mind when thinking about yielding at stress concentrations are the sharpness and hence gradient of the stress concentration—sometimes termed the constraint of the system—the average or nominal strain in conjunction with the SCF and the yield properties of the weld zone and parent material if relevant.

Consider first of all the case of a hole in a plate (*Figure 4.55*). A small hole in a large plate corresponds to a localised 'highly constrained' system with local plasticity whereas a larger hole in a smaller plate corresponds to a less constrained system with widespread plasticity. In the case of *Figure 4.55a* the SCF is 3.0 and it will be assumed that the design stress or nominal stress in the plate is $\frac{2}{3}\sigma_Y$. The behaviour of the strain at the hole is shown schematically as the plate is loaded. At a load P_Y first yield occurs (as dictated by a yield criterion, in this case maximum stress $= \sigma_Y = P_Y \times$ SCF/area). Beyond P_Y the load/strain graph will be almost linear since the yielding will be localised due to the large bulk of surrounding elastic material. At another load P_s such that

$$\frac{P_s}{A} = \tfrac{2}{3}\sigma_Y, \quad \text{i.e.} \quad \frac{P_s}{A} \times \text{SCF} = 2\sigma_Y$$

the strain will have reached a point B. On unloading, the graph is

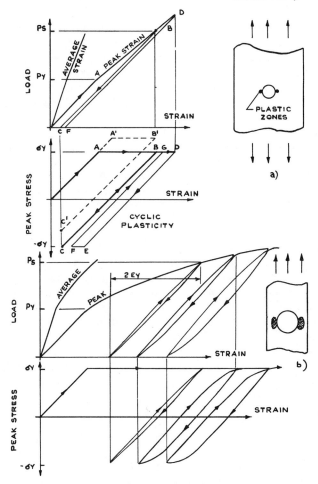

Figure 4.55

parallel to the loading line (the geometry has not significantly changed due to yielding) and at zero load the local strain is almost brought back to zero by the surrounding elastic material. The strain range on unloading is $2\varepsilon_Y$. Subsequent loads up to P_s produce completely elastic behaviour from C to B in *Figure 4.55a*.

The corresponding stress histories can be followed as shown, on the assumption of an elastic/perfectly plastic stress/strain law. The elastic behaviour after first yield is known as *shakedown* and emphasises that the material has shaken down to elastic behaviour. The shakedown load (P_s here) is the maximum load where the material will shake down. If the load is increased beyong P_s to D, say, it can be seen that plastic deformation will occur during each cycle of load. The behaviour on the loop FGDE sketched in *Figure 4.55a* is known as *reversed plasticity*. It is not difficult to appreciate that welds (or other details) which are allowed to exceed the shakedown condition will have a shortened cyclic life. The broken line AA'B'C' demonstrates that if the material adjacent to the SCF has an elevated yield point (as frequently is the case with welds) the effect is to raise the peak stresses and the shakedown limit.

In the second case, *Figure 4.55b*, the plastic zone is less constrained, the load/strain behaviour beyond the first yield load P_Y is less linear and a substantial amount of permanent deformation appears during the first cycle of loading. Shakedown will still occur and subsequent loading to P_s still only embraces a strain range of $2\varepsilon_Y$. In this type of situation when the load is increased beyond P_s, increments of permanent strain tend to be added during each cycle and one may feel intuitively that such a component will not last long. One would be correct. That type of behaviour has the descriptive term *incremental collapse*. By the above type of argument, the shakedown factor, defined by P_s/P_Y would always be 2.0 and the strain and stress ranges would always be $2\varepsilon_Y$ and $2\sigma_Y$ respectively during subsequent loading. While these statements are probably true for many components with highly constrained stress concentration factors, they are unfortunately not true in general.

Other features of shakedown behaviour

Shakedown factors are 2.0 only if the principal stress ratio remains constant during loading. This is difficult to be sure of in practice, but if the stress ratio alters, it tends to reduce the factor. A fuller but simple discussion of the shakedown concept is given in reference 16. Shakedown in bar structures is also discussed in reference 17.

It should also be noted that the shakedown load gives no indication of the permanent strain which may be generated in reaching the shakedown condition—it only ensures that subsequent loading will not induce strain ranges greater than $2\varepsilon_Y$. Initial permanent strains may be very large. It is also possible that collapse may occur before shakedown. For example the limit load for a simple beam is only 1.5 times the first-yield load (i.e. less than 2). In *Figure 4.55b* yielding may take place over the entire cross section at the hole before P_s is reached. Subsequent unloading and reloading to the maximum load reached could then be elastic, but the stress and strain ranges would then be less than $2\sigma_Y$ and $2\varepsilon_Y$ respectively. It is therefore sensible to redefine the shakedown load either as the collapse load, or as the shakedown value as given by the previous arguments.

The benefits of shakedown are therefore most often realised in situations where the SCF is moderate (say 2 to 8), and the constraint of the system is high enough to minimise initial permanent strain. Very high SCFs in highly constrained locations (with steep stress gradient) are probably better treated as if they were cracks (see Chapter 5). Events are changing rapidly in this field, and simplified methods of obtaining shakedown factors are emerging[18].

In the above treatment, the so-called Bauschinger effect has been ignored. If the 'reversed' yield strength is found to be different from the 'first load' yield strength, then the above arguments must be altered accordingly. Similarly it was assumed that the loading was in one direction. If reverse loading occurs of equal magnitude to the forward loading, then the shakedown load corresponds to first yield. A stress range of twice the yield strength is still available, but it now corresponds to the total load range.

Figure 4.56 is an attempt to summarise different situations which may arise at weld stress concentrations. Each case should be followed through in turn.

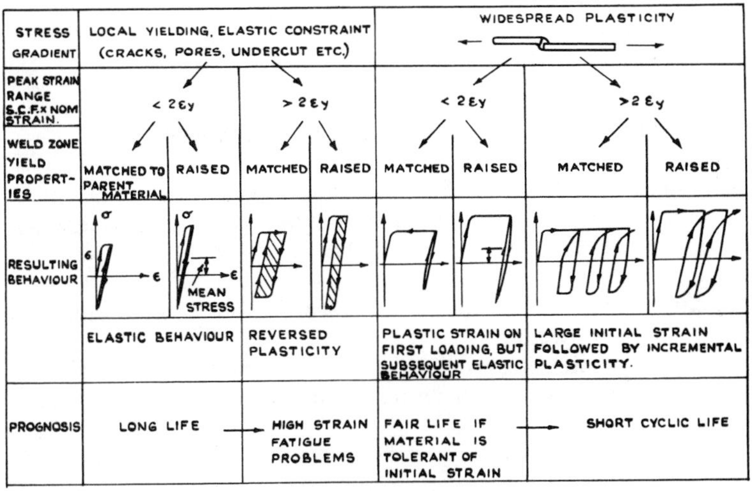

PLASTICITY AT NOTCHES

Figure 4.56

4.3.3 Residual stress

It will be readily seen that yielding at stress concentrations gives rise to residual stresses on unloading. Since residual stresses exist at zero load, they must be 'selfequilibrating', i.e. in equilibrium by themselves.

Thus if an area of tensile residual stress exists in a component, it must be balanced by a residual compressive stress *in the same direction* elsewhere.

Loading beyond the yield point is therefore one method of generating residual stress. Usually it engenders a favourable type of residual stress since the material is in a suitable state for future applications of load providing future loads are of the same type as the initial (proof test) load.

Residual stresses, however, are usually thought of in terms of the unfavourable type which are difficult to evaluate and normally arise due to welding or fabrication.

Residual stress due to welding

Residual stresses due to welding arise from the differential heating of the components by the weld heat source and the resulting non-uniform plastic deformation. The development of residual stress was mentioned incidentally in Section 1.4 on distortion, but the reader may appreciate the simpler treatment which follows.

Generation of residual stress in longitudinal welds

A crude attempt at describing the stage-by-stage development of residual stresses in the making of a longitudinal butt weld between two plates is shown in *Figure 4.57*. The stages are as follows:

Stage 1 Plate and weld (hypothetical) before commencing and unrestrained

Stage 2 Weld being produced at temperature but plate considered as unstrained. Thermal strain only is produced with zero stress

Stage 3 Weld and plate considered connected but the weld is still hot. Note the strain at 3 must be the same for weld and plate

The final cooling has been separated into two stages for convenience.

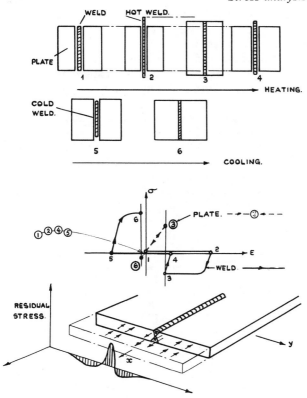

Figure 4.57   Development of residual stress

Stage 4   Separation of weld and plate. (Not the same as stage 2 since plastic deformation has taken place at stage 3)

Stage 5   Cooling of weld

Stage 6   Final state

By following the stress and strain history of the weld and plate, it is seen that as a result of heating the central strip representing the weld (it need not even be a weld), a residual tensile stress of approximately yield point magnitude is developed over a narrow zone at the centre. A considerably smaller compressive stress is developed on either side over a relatively wide zone. Residual welding or heating stresses, like those already mentioned due to loading, require to be selfequilibrating. A number of important points should be borne in mind.

1 If attempts are made to minimise distortion by 'balanced' welding then the residual stresses may also be balanced and in some cases their magnitude may be larger. The effect depends somewhat on the configuration in question.

2 A weld material which has a high yield strength, especially at elevated temperature, will allow high residual stresses to be developed.

3 The selfequilibrating feature of residual stress is important when hypothesising possible states of residual stress and can be useful in confirming the validity of measured residual stresses.

4 If plates are thick relative to the heat flow pattern (*Figure 4.58*) then residual stress patterns of significant magnitude can develop across the thickness but in the y direction of *Figure 4.57* (see also references 19 and 20). For even thicker plates, it is possible to generate significant residual stresses in the z direction

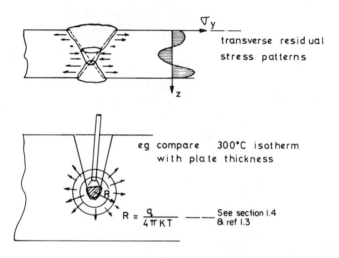

Figure 4.58

normal to the plate surface. This latter case is especially likely in the case of a plug-type repair in a thick plate. Because of the selfequilibrating condition above, the maximum stress is liable to be in the centre of the plate thickness and close to the weld.

5 It might be helpful here to distinguish between residual stresses developed in the manner described in *Figure 4.57* and so-called 'reaction' stresses generated by interaction of supports or other members in a welded structure. These latter residual stresses

would be better described as 'suppression of distortion' stresses. The former originally described are for a plate free to move or distort in whichever manner it wishes. Consider, for example, the two cases in *Figure 4.59* of stiffeners being welded to a long flat plate. In the first case residual stresses will be present at the junction and the distortion might be as shown. In the second case if the stiffeners are constrained to be parallel, the distortion is suppressed and altered. In addition to welding residual stress,

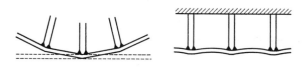

Figure 4.59

there are stresses (probably of different sign) due to the restraint. This type of situation has obvious implications for plate structures, ships, box girders, etc.

Thermal stress relief

Relief of residual stress by thermal means is perhaps the most common and most efficient. It is a plasticity/creep process. By elevating the temperature the yield stress is lowered, creep occurs rapidly and redistributes the stresses via creep deformation until hopefully they reduce to a low level. Usually, however, stress relief times are based on uniaxial relaxation test data. Unfortunately this can be misleading in some cases. Where residual stress systems are essentially uniaxial such as the *thin* plate of *Figure 4.57*, thermal stress relief on this basis can be expected to be effective. Where multiaxial residual stresses exist such as in 4 above, the situation is not so clear. Both plasticity and creep processes are generally agreed to depend on the equivalent stress (see page 181). Since this is unaffected by hydrostatic stress, one might expect that triaxial stress systems would be slower to relax than uniaxial ones, the slowness depending on the triaxiality. This area is not yet quantified enough to give general guidance but care is obviously required. Normal thermal stress relief processes based on uniaxial tests are unlikely to be adequate for complex thick geometrical features involving highly constrained welds.

Thermomechanical stress relief

Some other methods of stress relief aim at using thermal expansion to provide the mechanical forces required to set up other residual stress systems to counteract the original. They have the advantage of being able to be applied in situ but the disadvantage of needing some expertise coupled with some knowledge of mechanics for efficient operation. Essentially they depend upon a thermomechanical interaction to counteract existing residual stress. Consider the process of applying two bands of heat to either side of a longitudinal weld. The heating cycle produces residual stresses in exactly the same manner as described in *Figure 4.57*, the positions of the heat bands being chosen such that the new residual stresses counteract and cancel the original ones due to welding as shown in *Figure 4.60*. Such a process sensibly applied, while not expected to eliminate residual stresses, can reduce them to a low level.

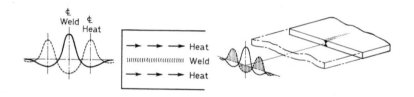

Figure 4.60

Spot heating of isolated spots (perhaps after repair or at the end of fillet welds, etc.) is a similar situation but rather more difficult to apply efficiently than band heating. By analogous arguments the heated spot deforms plastically initially before the surrounding area gets a chance to heat up. On removal of heat the spot cools quicker to a level where it can develop elastic stress again.

Subsequent slower cooling of the surrounding mass of material generates a radially symmetric residual stress system, which is biaxial tensile at the centre of the spot, with a circumferential compressive stress at some distance from the centre. If this technique is to be used, a precise knowledge of the position of the compressive region will be vital, as the residual stresses might be made worse rather than better. Some advice is given in reference 10, Chapter 6.

Mechanical stress relief

Purely mechanical stress relief can also be applied provided sufficient ductility is available to provide the necessary deformation. That is an easy statement to make but is rather difficult to quantify in practice. To examine the behaviour of a component containing initial stress during loading, it is convenient to imagine that the plate in *Figure 4.57* is loaded in tension along the *x* direction by gripping the ends in such a way that the whole width of the plate strains uniformly in that direction, i.e. the strain in the *x* direction is the same for weld and plate. The complex stress distribution shown in *Figure 4.57* will be approximated by the simplified distribution shown in the idealised diagrams on the right-hand side of *Figure 4.61*. Note that at zero load $\sigma_A \times$ Area *A* must be equal to $\sigma_B \times$ Area *B* for equilibrium. The stress–strain behaviour is developed in *Figure 4.61* for pieces of material initially at stress levels corresponding to A, B and C. The corresponding load/strain behaviour for the component is shown.

When the load is raised to position 1 and removed (1′) the residual stress at A is reduced; so is the compressive residual stress at B but a tensile residual stress has been generated at C. Note that the strain is the same for each point across the plate. Similarly, loading to position 2 and unloading (2′) changes the stresses at ABC to the values shown. At load point 3 all the plate has reached tensile yield and the final stresses on unloading must obviously be zero since the plate is now acting like a large tensile specimen. Any attempt to increase the load to 4 say, does not change the stress picture but simply adds strain.

One can see then that applying even a small load will reduce the value of the peak residual stress and subsequent behaviour up to that load will be elastic. At the load point 3 the residual stresses have been completely removed. This technique has a certain amount of practical use and is called Mechanical Stress Relief. In the case of a more complicated component having a stress concentration, the initial disadvantageous residual stresses are still removed but continued loading would then produce advantageous (negative) residual stress via the shakedown process already described. The following observations can thus be made:

1 Components having residual stresses of yield point magnitude do not collapse on application of load.
2 Significant plastic deformation will be necessary to ensure that future applications of the load will produce only elastic behaviour. Application of any load (once) will ensure this elastic

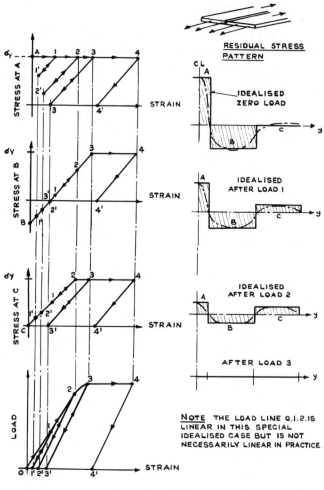

Figure 4.61

behaviour subject of course to the limitation that the shakedown load must not be exceeded.

3 The condition necessary to completely remove the residual stress is not easy to determine in general. In the simplified case of *Figure 4.61* all residual stresses have been removed when a strain of twice the yield strain is reached—point 3 unloads to 3'. In more complex components it is unlikely that each part of the structure will be strained to that level even at the highest

envisaged loads. Consequently *much higher strains* may be required at some locations before all the residual stresses are removed. It depends on the location and orientation of the residual stress compared with the stress orientation and distribution produced by the loading. Where these conflict the residual ones will be removed as described above and/or, via the shakedown process, may be replaced by ones of opposite sign which will be favourable with respect to the load being applied.

4 Subsequent loads should be the same type as the initial one or the process will be repeated with yet more plastic deformation. Thus test loads should be representative of service loading otherwise unnecessary and perhaps unfavourable deformation and stress cycles occur.

Lest the fortuitous behaviour described above lulls one into a false sense of complacency, it is as well to remember that the effect of residual stress on fatigue and fast fracture has not been considered (see Chapter 5). Also, it should not automatically be assumed that the material, at A in *Figure 4.61*, for example, is capable of being strained to the extent necessary in order to alter the residual stress. It has already been subject to a considerable amount of strain during the thermal cycle(s) shown in *Figure 4.57* and is unlikely to be in a healthy condition. It may crack during loading.

Measurement of residual stress is difficult and one should be wary of accepting measured values without being convinced of their validity. Several of the commercial instruments available require superlative skill to be at all accurate. Reference 21 gives several investigations into residual stress measurements. Perhaps the most reliable simple type of residual stress measurement is by attaching a strain gauge rosette to the surface and trepanning a groove around the location. It is sometimes not necessary to remove a plug of material completely but much care is required in the operation and in the interpretation of the results. Direct calibration of the method is essential rather than arguments regarding accuracy of instruments or observations and much larger scatter than is normal in strain gauge work must be expected.

REFERENCES

1. BENHAM, P. P., *Elementary mechanics of solids*, Pergamon (1965)
2. CRANDALL, S. M., and DAHL, N. C., *An introduction to the mechanics of solids*, McGraw-Hill (1959)

3. DRUCKER, D. C., *Introduction to the mechanics of deformable solids*, McGraw-Hill (1967)
4. ROARK, R. J., *Formulas for stress and strain*, McGraw-Hill
5. TIMOSHENKO, S. P., and GERE, J. M., *Mechanics of Materials*, Van Nostrand (1972)
6. DOVE and ADAMS, *Experimental Stress Analysis and Motion Measurement*, Merrill
7. DRUCKER, D. C., 'Thoughts on the present and future interrelation of theoretical and experimental mechanics', *Exp. Mechs.*, 97 (1967)
8. *Steel Girder Bridges*, BS 153, British Standards Institution, London
9. MOCANU, D., and BUGA, M., 'Distribution des contraintes le long des Cordons Lateraux de Sondures et dans les Toles', *Proc. Instn Mech. E.* (1970)
10. PETERSON, R. E., *Stress concentration factors*, Wiley, N.Y. (1953)
11. GUIAUX, P., 'Static tests on aluminium alloy welded truss joints', *British Welding J.*, **14**, No. 11 (Nov. 1967)
12. KOENIGSBERGER, F., *Design for welding in Mechanical Engineering* (1948)
13. ROSE and THOMPSON, BWRA Report D3/17/60
14. *Handbook for Welding Design*, Vol. 1, Pitman (1967)
15. COKER and FILON, *Photoelasticity*, Cambridge (1957)
16. FINDLAY, G. E., and SPENCE, J., 'Applying the shakedown concept to pressure vessel design', *The Engineer* (12th July 1968)
17. PRAGER, W., *An introduction to Plasticity*, Addison Wesley (1959)
18. MCFARLANE, W. A., and FINDLAY, G. E., 'A simple technique for calculating shakedown loads in pressure vessels', *Proc. Instn Mech. E.*, **186**, No. 4 (1972)
19. HALL, W. J., KIHARA, H., SOETE, W., and WELLS, A. A., *Brittle fracture of welded plate*, Prentice Hall, 270 (1967)
20. OATES, G., and PRICE, A. T., 'Appl. of fracture mechanics to boiler drum nozzle repairs', Practical application of fracture mechanics to pressure vessel technology. *Proc. Instn Mech. E.* (1971)
21. 'Residual Stresses', Welding Research Council Bulletin No. 121 (Apr. 1967)
22. KENYON, N., *et al.*, 'Fatigue strength of welded joints in structural steels', *British Welding J.*, **13**, No. 3 (Mar. 1966)
23. PANKHURST, R. C., *Dimensional Analysis and Scale Factors*, The Institute of Physics and the Physical Society Monographs for students
24. NOVOZHILOV, V. V., *Thin Shell Theory*, P. Noordhof Ltd (1964)
25. TIMOSHENKO, S. and WOINOWSKY-KRIEGER, S., *Theory of Plates and Shells*, 2nd edn, McGraw-Hill (1959)
26. TIMOSHENKO, S. and GERE, J. M., *Theory of Elastic Stability*, 2nd edn, McGraw-Hill (1961)
27. TIMOSHENKO, S. and GOODIER, J. N., *Theory of Elasticity*, McGraw-Hill (1951)
28. HUTTON, G. and ROSTRON, M., *Computer Programs for the Building Industry 1979*, Architectural Press, London (1979)

BIBLIOGRAPHY

'Stress Concentrations', Engineering Sciences Data Unit—Items 65004B, 66004, 66035, 67023A, 69021A, 251 Regent St., London W1R 7AD

Chapter 5

Fracture

5.1 The theoretical basis of fracture mechanics

5.1.1 Introduction

Engineering structures of all kinds, including welded structures, have been breaking unexpectedly in service for as far back as records are available, and until comparatively recently effective methods for assessment and control of fracture were simply not available. Development of the new discipline called 'fracture mechanics' has proceeded by rather subtle stages and as a consequence, the underlying basis of the various techniques which are now used may not always be apparent. In this chapter, therefore, the aim is as much to expose the scope and background of fracture mechanics as to describe the techniques used.

The nineteenth-century mechanician, Gustav Kirchoff, once said: 'the task of Mechanics is to *describe* observable movements in as complete and simple a manner as possible . . .', and we have seen in Chapter 4 how the mechanics-based concepts of stress, elastic deformation and buckling instability can be used to assess and predict important aspects of the behaviour of loaded structures. Unfortunately, these concepts have not always proved to be adequate in describing certain kinds of structural *failure*, or to put it another way, structures have often collapsed although the *stresses* calculated according to the best methods available were shown to be 'acceptable'. These experiences frequently embarrassed practical men and theoreticians alike, but little progress was made until the role of *pre-existing cracks* came to be recognised.

The essential assumption which underlies fracture mechanics assessment therefore, is that crack-like defects are almost always present in fabricated structures and that therefore some kind of mechanical description of the crack-tip region should be sought in order to assist understanding of the tendency to crack propagation. ('Fracture mechanics' is usually understood to imply '*sharp-crack* fracture mechanics'.) Normally, the crack is assumed to be much larger than typical microstructural features and can therefore be

217

considered to be located in the usual isotropic continuum which exhibits idealised constitutive properties. Of course, processes other than crack propagation may be involved in fractures. For example, fatigue crack initiation (see Chapter 6) or the formation of microstructural defects during welding may precede the appearance of a macroscopic crack, or the growth of voids during creep or plastic flow may lead directly to separation and failure. These processes are not normally treated by fracture mechanics as defined here.

The essential tools in fracture mechanics technology are *cracked-body solutions* (in essence stress analyses) which are used to *describe* important features of the crack-tip region. These solutions must of course reflect the important features of real problems if they are to be useful and in particular, care is taken to describe the loading, the geometry of the structure including the crack size and shape, the constitutive properties of the material and the three-dimensional character of the crack-tip mechanical state (see *Figure 5.1*). The historical picture of fracture mechanics portrays development from the treatment of simply loaded shapes made of linear elastic material, to the present day when complex shapes made of more realistic elastic/plastic materials can be described.

Fracture mechanics also makes it possible to *compare* the behaviour of different cracked bodies made of the same material and this idea is central to its application in engineering practice. In particular one can test a cracked specimen in the laboratory and by relating the crack-tip stress field at fracture to the test load, derive a *critical* value of fracture parameter appropriate to the given material and environment. This value can then be used to determine the combinations of load and crack size which will induce fracture in different structures of practical interest (see *Figure 5.2*). (This philosophy has its parallels in Chapter 4 where it can be seen that a *yield strength* determined via a simple tensile test, may be used to predict yielding in a beam or a pressure vessel.) The attractive feature of this idea is that it is not essential to understand the precise *causes* of fracture at the microscopic level. One only requires an assurance that identical crack-tip mechanical states have been produced in the test specimen and in the real structure.

5.1.2 Crack-tip stress analysis in linear elastic structures

The earliest relevant solution for a cracked body was given by Inglis[1]. It gives a complete and *exact* (see Section 4.1.10) description of the stresses in a linear elastic plate pierced by a smooth elliptical hole and

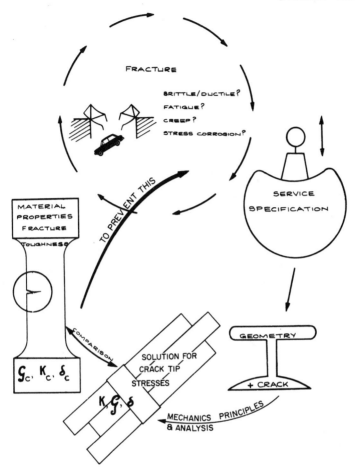

Figure 5.1

loaded uniformly at an infinite distance from the hole (*Figure 5.3*). The stress pattern adjacent to the hole is difficult to describe or to visualise in simple terms, but it is easy to identify the *maximum* stress which occurs at the 'sharp' end of the ellipse, normal to the major axis and is given by

$$\sigma_{max} = \sigma(2a/b)$$

As Inglis pointed out, the solution is valid for *any* elliptical shape, and if the minor axis *b* is reduced relative to the major *a*, the solution for a sharp crack is effectively obtained, whereupon the maximum stress

Fracture mechanics solutions must recognise these
features of practical crack problems.

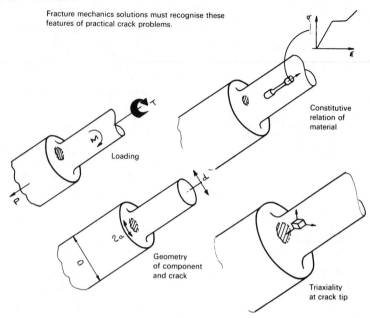

Constitutive
relation of
material

Loading

Geometry
of component
and crack

Triaxiality
at crack tip

Figure 5.2

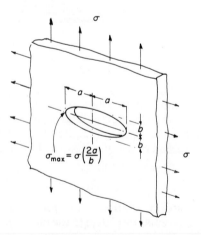

$$\sigma_{max} = \sigma\left(\frac{2a}{b}\right)$$

Figure 5.3   Inglis' solution

tends to infinity. Inglis first described this solution during a conference at the Institute of Naval Architects in 1913 and it is clear from the printed discussion that some of the audience were quite uncertain as to the implications and relevance of the results. How could one relate this infinite stress to the yield or ultimate strength of the material? Did it mean that all cracks would propagate, regardless of stress or crack size? Some suggested that the intervention of yielding at the crack tip would invalidate the analysis which was therefore of little practical value. Perhaps for these reasons, the solution was set aside for several years as far as application to cracks was concerned and little was made of Inglis' important conclusion that if one assumes that all cracks have more or less the same crack-tip shape, the maximum stress can then be expressed as

$$\sigma_{max} \propto \sigma\sqrt{a}$$

—a relationship which gives a very good impression of the relative significance of applied stress and crack length.

Eventually in 1921, Griffith[2] used Inglis' solution to derive an alternative crack-tip description in terms of elastic energy variation in the plate during a small increase in crack size. This turned out to be a finite quantity, despite the infinite crack-tip stress and therefore avoided one of the notional difficulties in relation to the use of Inglis' solution. However, although Griffith's approach is of considerable academic and historical interest, it is less frequently used in engineering practice today and will not be pursued further here, except to note that Griffith also showed experimentally that the linear elastic model was indeed relevant to real fractures, in that cracked glass specimens broke under constant conditions determined by the product of stress and the square root of crack length as foreshadowed in the equation above.

Theoretical and experimental progress was thereafter very slow and although several solutions for various notched and cracked body shapes were produced in the period straddling the Second World War, it was not at all clear how these could be utilised. The light did not seem to dawn until in the nineteen-fifties Irwin[3] and Williams[4] recognised independently that the stresses very close to crack tips follow a characteristic pattern irrespective of the body shape or loading. Before describing this pattern, it is convenient to adopt Irwin's 'mode' classification of cracked bodies according to the loading direction relative to the crack plane (*Figure 5.4*). In terms of a polar coordinate system centred on the crack tip (see *Figure 5.5*) the near-crack-tip stresses ($r \ll a$) are given by

$$
\begin{matrix} \sigma_y \\ \\ \sigma_x \\ \\ \sigma_{xy} \end{matrix}
= \frac{K_I}{\sqrt{(2\pi r)}} \cos\left(\frac{\theta}{2}\right)
\begin{bmatrix} 1 + \sin\left(\frac{\theta}{2}\right)\sin\left(\frac{3\theta}{2}\right) \\[2mm] 1 - \sin\left(\frac{\theta}{2}\right)\sin\left(\frac{3\theta}{2}\right) \\[2mm] \sin\left(\frac{\theta}{2}\right)\cos\left(\frac{3\theta}{2}\right) \end{bmatrix}
$$

or

$$
\begin{matrix} \sigma_y \\ \\ \sigma_x \\ \\ \tau_{xy} \end{matrix}
= \frac{K_{II}}{\sqrt{(2\pi r)}}
\begin{bmatrix} \sin\left(\frac{\theta}{2}\right)\cos\left(\frac{\theta}{2}\right)\cos\left(\frac{3\theta}{2}\right) \\[2mm] -\sin\left(\frac{\theta}{2}\right)\left[2 + \cos\left(\frac{\theta}{2}\right)\cos\left(\frac{3\theta}{2}\right)\right] \\[2mm] \cos\left(\frac{\theta}{2}\right)\left[1 - \sin\left(\frac{\theta}{2}\right)\sin\left(\frac{3\theta}{2}\right)\right] \end{bmatrix}
$$

or

$$
\begin{matrix} \tau_{xz} \\ \\ \tau_{yz} \end{matrix}
= \frac{K_{III}}{\sqrt{(2\pi r)}}
\begin{bmatrix} \sin\left(\frac{\theta}{2}\right) \\[2mm] \cos\left(\frac{\theta}{2}\right) \end{bmatrix}
$$

for the three modes, I, II and III.

K, which is called the 'crack-tip stress-field-intensity factor' is a function of the loading, the crack size and other geometrical features, and is given a subscript I, II, or III to indicate the mode. It is *not* a stress concentration factor (see Section 4.3.1) but gives the 'amplitude' of the characteristic stress pattern through the equations above. For example, in the classical infinite plate geometry loaded in mode I $K_I = \sigma\sqrt{\pi a}$, whereas in the case of a slender beam which is being wedged open by a force P/unit thickness, $K_I = Pa2\sqrt{3}/c^{3/2}$ (see *Figure 5.6b*). Despite this difference in configuration between the two cases in *Figure 5.6*, which will undoubtedly lead to different stresses in the body as a whole, the stress patterns near to the crack tip in each case will be *identical* if the magnitude of K_I is the same for each body. This conclusion emerges clearly from the equations above which show that for a given mode, the stress at a given point relative to the crack tip depends only on the magnitude of K. As the stress fields are identical, we can therefore confidently argue that the tendency to crack propagation is identical.

In practice therefore, if one of these bodies happens to be a laboratory test specimen, we can use the fracture load and crack length to determine a 'toughness property' designated K_{Ic} ('c' for critical) which can then be used to predict critical combinations of

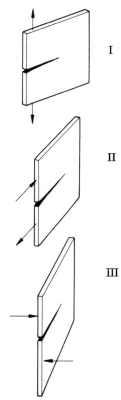

Figure 5.4 Loading modes

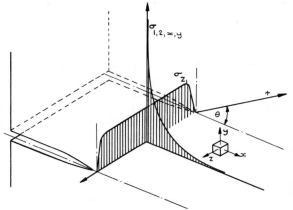

Figure 5.5 Linear elastic stress distribution

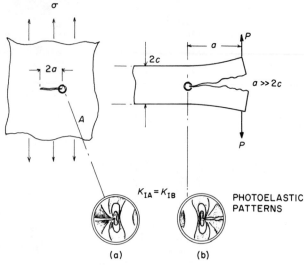

Figure 5.6   *Crack-tip stress patterns in bodies loaded in an unlike manner*

load and crack size in other practical cases where the same material and environment applies. Note that the *actual values* of crack-tip stress at fracture are of no particular interest, as one is not attempting to relate such stresses to microstructural strength properties or to the tensile strength properties exhibited by an uncracked specimen. Also, remember that no direct comparison can be drawn between mode I fracture and modes II or III fractures—as the angular dependence of stress is different for all three modes, the stress fields will look different, and therefore we should not expect similar behaviour.

The liberating aspect of the invention of K was that it was no longer necessary to ascertain the complete and bewildering detail of overall stress distributions in cracked bodies, as the important features could be characterised by the single parameter K. This opened up the field for the development of catalogues of solutions giving the value of K for different geometries (e.g. references 5 to 9).

A number of the more useful results of K derivations have been included in *Figures 5.7* to *5.9* which show various shapes of cracked body. It is convenient to present the results in the form of an adjustment to the infinite plate solution, i.e.

$$K_I = \sigma\sqrt{\pi a F}$$

where F is a function of various dimensions of the body, ratioed to the crack length. In many cases a formula can be constructed for this

Case 1 — CENTRAL INTERNAL CRACK-TENSION-LENGTH > 4W

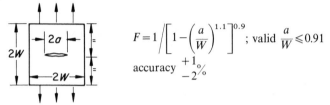

$$F = 1 \bigg/ \left[1 - \left(\frac{a}{W} \right)^{1.8} \right]^{1.08} \; ; \text{valid } \frac{a}{W} \leqslant 0.91$$

accuracy $\pm 1\%$

(for eccentrically located cracks see case 12)

Case 2 — CENTRAL INTERNAL CRACK-TENSION-SQUARE

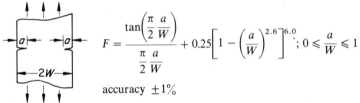

$$F = 1 \bigg/ \left[1 - \left(\frac{a}{W} \right)^{1.1} \right]^{0.9} \; ; \text{valid } \frac{a}{W} \leqslant 0.91$$

accuracy $^{+1}_{-2}\%$

Case 3 — DOUBLE EDGE CRACK-TENSION-INFINITE LENGTH

$$F = \frac{\tan\left(\dfrac{\pi}{2} \dfrac{a}{W} \right)}{\dfrac{\pi}{2} \dfrac{a}{W}} + 0.25 \left[1 - \left(\frac{a}{W} \right)^{2.6} \right]^{6.0} \; ; 0 \leqslant \frac{a}{W} \leqslant 1$$

accuracy $\pm 1\%$

Case 4 — SINGLE EDGE CRACK-TENSION-INFINITE LENGTH

$$F = 1.25 \bigg/ \left[1 - 0.7 \left(\frac{a}{W} \right)^{1.5} \right]^{6.5} \; ; \frac{a}{W} \leqslant 0.65$$

accuracy $^{+1.4}_{-0.6}\%$

Figure 5.7

adjustment as shown in the figures, or curves can be used to show the relations as in *Figures 5.10–5.12* and in Rooke and Cartwright[7].

This information will of course require detailed study but it is worth pointing out a number of useful general points. Firstly, edge cracks have a 12% higher K factor than internal cracks, other things being equal (cf. cases 1, 2, 8, 9, with 3, 4, 5, 6, and 7 for a/W tending to zero. Also compare cases 10 and 11). To be consistent with the infinite plate solution, the stress used in the formula is normally the *gross*

226

Case 5 — SINGLE EDGE CRACK-PURE BENDING-INFINITE LENGTH

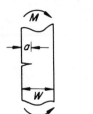

$$F = \left(1.25 \Big/ \left[1 - \left(\frac{a}{W}\right)^{1.82}\right]^{2.57}\right) - \sin\left(\frac{\pi a}{2W}\right); \ \frac{a}{W} \leqslant 0.7$$

accuracy $\begin{smallmatrix}+2\\-1\end{smallmatrix}\%\left(K_I = \frac{M}{\frac{1}{6}BW^2}(\pi a F_5)^{1/2}\right)$

K_5 can be added to K_4 to give tension plus bending
B is plate thickness

Case 6 — SINGLE EDGE CRACK-3-POINT BENDING-SPAN$=4W$

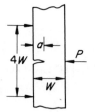

$$F = \left(1.19 \Big/ \left[1 - \left(\frac{a}{W}\right)^{1.4}\right]^{1.87}\right) - 0.9 \sin\left(3\frac{a}{W}\right); \ \frac{a}{W} < 0.7$$

accuracy $\begin{smallmatrix}+2\\-1\end{smallmatrix}\%\left(K_I = \frac{PW}{\frac{1}{6}BW^2}(\pi a F_6)^{1/2}\right)$

B is plate thickness

Case 7 — CIRCUMFERENTIAL CRACK-TENSION-INFINITELY
      LONG CYLINDRICAL BAR

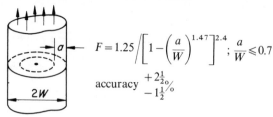

$$F = 1.25 \Big/ \left[1 - \left(\frac{a}{W}\right)^{1.47}\right]^{2.4}; \ \frac{a}{W} \leqslant 0.7$$

accuracy $\begin{smallmatrix}+2\frac{1}{2}\\-1\frac{1}{2}\end{smallmatrix}\%$

Case 8 — SINGLE CRACK AT EDGE OF HOLE UNIAXIAL TENSION

$$F = \left[1 - \sin\left(\pi\frac{r}{a}\right)\right]\left[1 - \left(\frac{r}{a}\right)^{16.6}\right]^{0.9}$$
$$+ \left[\sin\left(\pi\frac{r}{a}\right)\right]1.45\left(\frac{r}{a}\right)^{1/3}; \ 0 \leqslant \frac{r}{a} \leqslant 1$$

accuracy $\begin{smallmatrix}+1\frac{1}{2}\\-3\end{smallmatrix}\%$

Figure 5.8

Case 9 — DOUBLE CRACK AT EDGES OF CIRCULAR HOLE-UNIAXIAL TENSION

$$F = \left[1 - \sin\left(\frac{\pi r}{a}\right)\right]\left[1 - \left(\frac{r}{a}\right)^5\right]^{0.78}$$

$$+ \left[\sin\left(\frac{\pi r}{a}\right)\right]1.23\left(\frac{r}{a}\right)^{0.19}; \quad 0 \leqslant \frac{r}{a} \leqslant 1$$

accuracy $\begin{array}{c}+3\\-2\end{array}\%$

Case 10 — ELLIPTICAL PLAN BURIED CRACK-TENSION

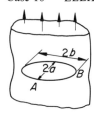

$$F_{A(max)} = \left(1 - 0.619\frac{a}{b}\right); \quad 0 \leqslant \frac{a}{b} \leqslant 1$$

$$F_{B(min)} = \left(\frac{a}{b}\right)\left(1 - 0.619\frac{a}{b}\right); \quad (K_1 = \sigma\sqrt{\pi b F_B})$$

accuracy $\begin{array}{c}+1\frac{1}{2}\\-2\frac{1}{2}\end{array}\%$

Case 11 — SEMI-ELLIPTICAL PLAN SURFACE CRACK-TENSION

$$F_{A(max)} = \left[1 + 0.12\left(1 - \frac{a}{b}\right)\right]^2\left[1 - 0.619\frac{a}{b}\right]; \quad 0 \leqslant \frac{a}{b} \leqslant 1$$

$$F_{B(min)} = \left[1 + 0.12\left(1 - \frac{a}{b}\right)\right]^2\left(\frac{a}{b}\right)^2\left[1 - 0.619\frac{a}{b}\right]$$

accuracy as case 10

Case 12 — ECCENTRIC INTERNAL CRACK-TENSION-LENGTH $> 4W$

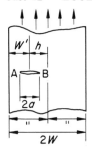

$$\lambda = \frac{a}{W'} \leqslant 0.9; \quad \varepsilon = \frac{h}{W} < 1.0$$

$$\theta_1 = \pi\frac{\lambda}{2}(1 - \varepsilon); \quad \theta_2 = \pi\left[\frac{\lambda}{2}(1 - \varepsilon) \pm \varepsilon\right]$$

(positive sign in θ_2 for left-hand end of crack (A)
negative sign for right-hand end (B))

$$F_{A,B} = \frac{1}{[1 - \lambda^{1.8}]^{1.08}}\left[\frac{2(1 - \cos\theta_1)(1 + \cos\theta_2)}{\theta_1\sin\theta_1(\cos\theta_1 + \cos\theta_2)}\right]\frac{\frac{\pi}{2}\lambda}{\tan\left(\frac{\pi}{2}\lambda\right)}$$

accuracy: mostly better than $\pm 3\%$ if $\lambda < 0.7$ *and* $\varepsilon < 0.9$

otherwise $F_A \begin{array}{c}+13\\-8\end{array}\%$

$$F_B \begin{array}{c}+16\\-10\end{array}\%$$

Figure 5.9

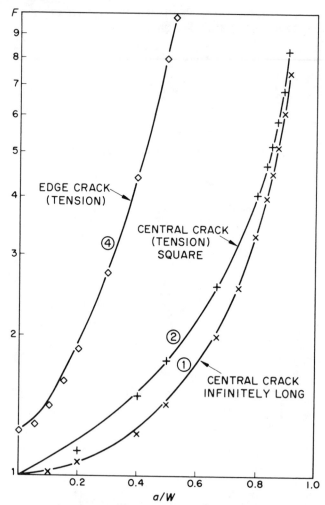

Figure 5.10a   Linear crack solutions (see Figure 5.7) (data source see reference 38)

stress (i.e. neglecting the reduction in load-carrying cross-section) and the sharp increase in K which occurs as the crack occupies more of the cross-section can be shown to be identical to the increase seen in the equivalent stress-concentration factor tables for finite-width sharp notches (reference 10 in Chapter 4)—indeed F factors can be derived from stress-concentration solutions and vice versa. Cases 8 and 9 for cracks growing out from holes (graphed in *Figure 5.11*) show that an edge crack which is greater than 10% of the hole size should be treated as

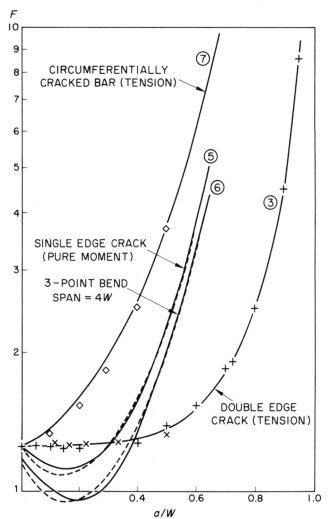

Figure 5.10b Linear crack solutions (see Figure 5.8) (data source see reference 38)

if it were part of a much longer crack which includes the hole diameter. Cracks smaller than this can be treated as if they were situated in a semi-infinite plate where the stress concentration due to the hole elevates the gross stress by a factor of 3 or whatever is appropriate. This finding has considerable relevance to the failure of Comet I aircraft where apparently 'short' cracks growing outwards from window openings had a much greater effective length than might have been realised. Finally, it

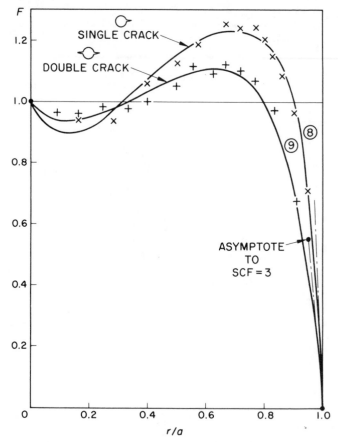

Figure 5.11   Linear crack solutions (see Figures 5.8 and 5.9)

should be noted that in the three-dimensionally-described elliptical cracks of cases 10 and 11, the stress-field-intensity factor varies along the crack border, the maximum value of K being at the ends of the *minor* axis.

Although a wealth of solutions for K exist, such is the nature of engineering practice that one always seems to have to assess problems for which no immediately applicable formulae have been published. In such cases a variety of techniques is available to combine or 'compound' solutions (reference 10). Most of these are based on *superposition* which is a valid procedure as the underlying models are linear (see Section 4.1.6). In particular, stress-field-intensity factors describing different loadings on the same geometry can be added

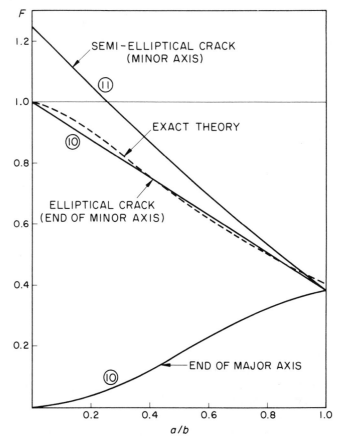

Figure 5.12   Linear crack solutions (see Figure 5.9)

algebraically, provided they are of the same mode. This trick is especially useful for evaluating the effect of residual welding stresses on internal cracks.

5.1.3 Plastic-zone-size correction

In the case of relatively high yield-strength, brittle materials, the long-standing objections to the use of linear elastic fracture mechanics (LEFM) for materials which yield (albeit at a high stress) was answered by the convincing evidence that yielding did not seem to matter in practice. That is, the materials which were commonly tested

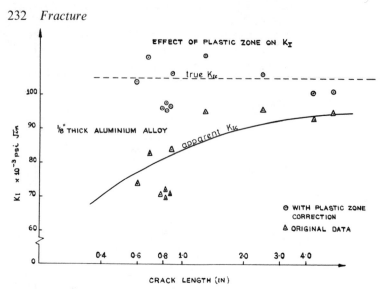

Figure 5.13

such as high-strength steel and aluminium alloys, glass and perspex, fractured at constant levels of K regardless of crack length or geometry. This conclusion also applied to medium-strength steels where unsympathetic welding and heat treatment procedures resulted in strain-hardening, deterioration in toughness and the retention of high residual stress fields (Wells[11]). However when the time came to apply LEFM to lower-yield-strength, higher-toughness situations, deficiencies began to appear, evidenced by apparent variations in K_c depending on specimen geometry. For example, in tests of an aircraft alloy, specimens with short initial crack sizes seemed to exhibit a lower toughness than specimens with long cracks (see full curve in *Figure 5.13*). This was a sure sign that the linear elastic description was no longer adequate and that some modification to recognise the development of crack-tip plasticity was called for.

Relaxation of the high crack-tip stress clearly throws even more load on the elastic material outside the plastic zone and a simple proposal was made by Wells and Irwin that an estimated plastic zone size (r_Y) should be added to the real crack length to form an effectively longer crack size which would be used for the calculation of K. Their first estimate of plastic zone size was made very simply by equating the crack-tip stress ahead of the crack to the uniaxial yield strength σ_Y so that

$$r_Y = \frac{1}{2\pi}\left(\frac{K}{\sigma_Y}\right)^2$$

and this form of correction appeared to work well in that a uniform level of K_c was obtained when the correction was included (upper broken line in *Figure 5.13*).

Some theoretical support for this simple adjustment was later derived from elastic/plastic solutions for mode III cracks (McLintock and Irwin[12]). These solutions were limited to r or $r_Y \ll a$, and showed the plastic zone to be circular of radius $K_{III}^2/(\tau_Y^2 2\pi)$ (see *Figure 5.14*).

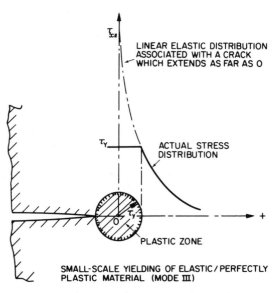

LINEAR ELASTIC DISTRIBUTION
ASSOCIATED WITH A CRACK
WHICH EXTENDS AS FAR AS O

ACTUAL STRESS
DISTRIBUTION

PLASTIC ZONE

SMALL-SCALE YIELDING OF ELASTIC / PERFECTLY
PLASTIC MATERIAL (MODE III)

Figure 5.14

The more significant feature was that the stress distribution *outside* the plastic zone was shown to be identical to that exhibited by an entirely linear elastic material where the crack is r_Y longer than the real crack. Overlooking the difference between modes I and III (the mode I plastic zone is not at all circular), this finding corroborated the suggestion that similar crack-tip mechanical states will be obtained if the crack is notionally lengthened, as differently cracked bodies loaded to the same value of 'plasticity-corrected' K will exhibit identical plastic zones and elastic stress patterns.

The mathematical expression of this idea therefore is that if $K_e = f(\sigma)f(a)$ gives the linear elastic stress intensity for a certain case, the 'plasticity-corrected' factor will be given by

$$K_p = f(\sigma)f(a + K_e^2/2\pi\sigma_Y^2)$$

To be logical, one should acknowledge that as the plastic zone size really depends on K_p rather than K_e, the correction should be repeated or 'iterated' until no further increase is obtained. In practice, if repeated iterations continue to increase K_p, the plastic zone size has probably become too large in relation to the crack size or the uncracked ligament for the procedure to be valid.

In the case of the infinite plate, the relation above becomes

$$K_p = K_e\left[1 + \frac{1}{2}\left(\frac{\sigma}{\sigma_Y}\right)^2\right]^{\frac{1}{2}}$$

This is called 'first-stage correction', and if iteration is carried through to its conclusion,

$$K_p = K_e\left[1 - \frac{1}{2}\left(\frac{\sigma}{\sigma_Y}\right)^2\right]^{-\frac{1}{2}}$$

For the infinite plate, comparison with a more rigorous solution to be given later, shows that the latter 'iterated' correction is remarkably accurate for $\sigma/\sigma_Y < 0.8$. However in finite-width geometries, large errors are developed for $F(a/W) > 2$ and $\sigma/\sigma_Y > 0.5$, and it is common in such cases to limit the correction procedure to the first stage.

5.1.4 Triaxiality and the plane strain–plane stress transition

By the end of the nineteen-fifties, some of the benefits of the fracture mechanics framework had begun to be realised, and in particular, resistance to crack propagation was being measured routinely on a sound basis. It had always been understood by some fabricators of welded structures that thick structures were for some reason significantly more prone to brittle catastrophic fracture than thin ones, and such effects which had hitherto been observed only dimly were exposed in a much clearer light by systematic K_c testing programmes. For example, tests carried out on aluminium alloy specimens over a range of thickness revealed a large and abrupt drop in toughness as thickness increased (*Figure 5.15*). This surprised many, as sudden transitions in toughness had previously been associated exclusively with ferrous materials. To understand these effects it is necessary to track the developing crack-tip stress state as the load increases from zero.

During the early stages when essentially linear elastic conditions

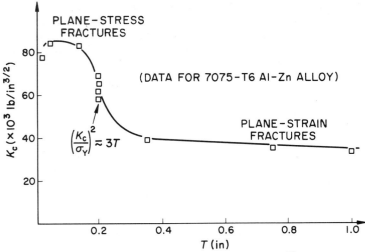

Figure 5.15   Toughness transition with thickness[13]

could be assumed, Irwin's equations show that the in-plane principal stresses ahead of the crack tip ($\theta = 0$) are equal and very high (the Mohr's circle reduces to a dot). The material in this intensely stressed zone would like to contract in the z or thickness direction due to the Poisson effect but is prevented from doing so by the surrounding elastic material which is at a low stress and therefore contracts very little. If, in the limit, it is assumed that the thickness strain at the crack tip is suppressed completely (a condition which is called *plane strain*) a tensile stress is developed in the thickness direction, viz.

$$\varepsilon_z = \frac{1}{E}(\sigma_z - v\sigma_x - v\sigma_y) = 0$$

(see Section 4.1.8) whereupon if $\sigma_x = \sigma_y$ and $v = 0.3$

$$\sigma_z = v(\sigma_x + \sigma_y) = 0.6\sigma_y$$

Thus a state of nearly equal triaxial or hydrostatic tensile stress is generated at the crack tip. This affects matters in two ways: first, yielding and plastic flow are greatly inhibited (see Section 4.1.11), the plastic zones stay very small and LEFM descriptions can be considered valid; and secondly, conditions which favour brittle, low-energy absorption, cleavage fracture of the crack-tip material are created. If fracture occurs at this stage it is called a 'plane-strain fracture' and as it is likely also to propagate in mode I, the measured toughness value is given the symbol 'K_{Ic}'.

As the applied load is increased, plastic zones form on either side of the crack plane ($\theta \simeq 90°$), where the yield criterion is more easily satisfied, and subsequently grow outwards and eventually forward as shown in *Figure 5.16a*. If the load is increased sufficiently, the plastic zones may reach a size where 45° shear flow completely through the thickness becomes possible (*Figure 5.16b*). For larger component

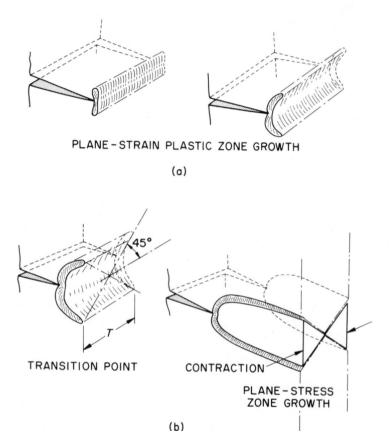

PLANE-STRAIN PLASTIC ZONE GROWTH

(a)

45°

TRANSITION POINT   CONTRACTION

PLANE-STRESS
ZONE GROWTH

(b)

Figure 5.16

thickness this can obviously only occur at larger plastic zone sizes and therefore at high levels of K. The material ahead of the crack *can* now contract in the z direction, independently of the adjacent elastic material; therefore the thickness stress drops, the yield criterion is more easily satisfied and for a minimal increase in load, the plastic zone suddenly enlarges. The process is self-stimulating as partial relaxation of the triaxial condition encourages plastic zone growth which stimulates further relaxation. With the beneficial reduction in triaxial stress and increase in shear flow, toughness usually improves substantially, which explains the abrupt transition shown in *Figure 5.15*.

The latter mechanical state is called '*plane stress*', as σ_z is considered to be negligible and a fracture occurring at this stage is designated a 'plane-stress' fracture. By convention, the mode I subscript is dropped from any quoted toughness values, although in practice the crack may still propagate on a plane normal to the loading.

The level of K at which transition to plane stress occurs is obviously of great importance but is not always easy to identify in a given case. From the simple pattern shown in *Figure 5.16b*, transition should be expected when the plastic zone diameter is equal to the thickness, that is when

$$\frac{1}{\pi}\left(\frac{K_I}{\sigma_Y}\right)^2 = T$$

However, in practice transition tends to occur at lower loads and the exact point is much affected by finite geometrical effects, work-hardening and loading configuration. Experimental evidence suggests that plane strain is unlikely to persist beyond $(K/\sigma_Y)^2 = T$ and for conservative plane-strain toughness testing procedures, an arbitrary limit of $(K/\sigma_Y)^2 < 0.4T$ has been adopted. Plane-strain conditions are also associated with buried cracks, as the plastic zones cannot reach a free surface which would allow relaxation, and situations where the uncracked ligament experiences bending tend to maintain plane strain as the compressive fibres will remain elastic to higher loads. Conversely if the plane-strain plastic zone penetrates to a free surface in a finite-width shape, transitions will occur prematurely.

The implication of these arguments is that a low-yield-strength material in a thin section will experience an earlier transition to a higher-toughness plane-stress condition. *Figure 5.17* gives an alternative assessment procedure for triaxiality in the classical infinite plate and this will be discussed in more detail later.

The triaxial state also has implications for the simple plastic zone size calculation. Under plane-strain conditions, the effective yield strength in the plastic zone is elevated (it is thought by a factor of about two) and logically this should be reflected in the calculation of zone size, viz.

$$r_Y = \frac{1}{2\pi}\left(\frac{K}{2\sigma_Y}\right)^2$$

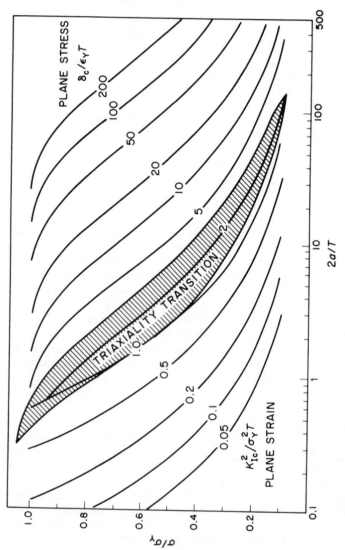

Figure 5.17 Assessment of triaxial constraint in infinite plate

which gives a zone size one-quarter of the nominal plane-stress zone size. In practice, this argument is very inconsistently treated in the literature—some researchers take account of it and others do not. Of course, the whole concept of plastic correction is rather rudimentary in any case, but whatever factor is used, a marked difference between plane strain and plane stress should be recognised.

5.1.5 Large plastic zones and the strip-yielding model

Linear elastic fracture mechanics characterisations, with plastic zone size adjustments if necessary, proved to be effective in dealing with situations where the nominal stress at fracture was a low proportion of the yield strength. In particular LEFM was shown to be apt for high yield strength/low toughness materials, especially in plane strain, and it was also applied successfully to plane-stress cases in lower-yield-strength materials where the plastic zone size was limited by thickness and/or low toughness to small proportions of the crack length or unfractured ligament.

There was, however, considerable incentive to extend the application of fracture mechanics to more ductile fracture modes. Pressure vessels, for example, are normally permitted to operate at high membrane stress levels (say two-thirds of the yield) and they usually contain stress concentrating features such as nozzles and end-closures which expose sizable volumes of material to multiples of the yield strain. Experience of full-scale high-strain fracture tests at the British Welding Research Association suggested that thick, welded pressure vessels could be at risk from cracks too small to be reliably discovered, especially where residual stress had not been removed. Several unexpected failures confirmed this prognosis.

The essential requirement for such extensions into the 'general yielding fracture mechanics' (GYFM) regime, as it became known, was an analytical model which included yielding behaviour, but was not restricted to small plastic zone size. In 1960, Dugdale[14] developed such a model which effectively simulated a particular pattern of cracked body yielding which he had observed in mild steel. This pattern is often found in thin sheets, particularly where the material shows a pronounced yield drop (see *Figure 4.3*). The plastic zones take on the appearance of 'candle flames' extending forward from the crack ends (see *Figure 5.18a*). This physical condition was realistically modelled by Dugdale in terms of a linear elastic composite solution where the narrow strip-yielding plastic zone was represented by a hypothetical extension of the real crack, the extended crack edges being restrained from opening up freely by yield-point-magnitude

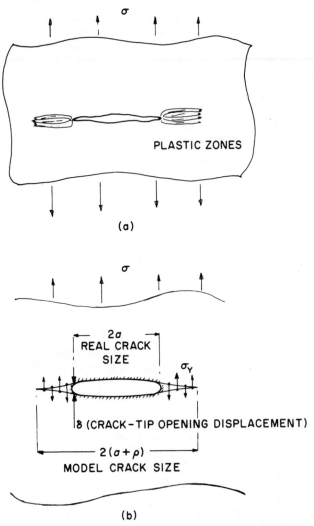

Figure 5.18   Cracked-body yielding: (a) experimental appearance; (b) mathematical model

forces applied normal to the crack plane (*Figure 5.18b*). For plane-stress conditions, the uniaxial yield strength is used, whereas when the model is pressed into service to describe plane strain, the effective yield strength might be doubled to allow for hydrostatic elevation.

Dugdale was primarily interested in the plastic zone length and showed that for the classical infinite plate case this is given by

$$\rho = a \left[\sec\left(\frac{\pi}{2} \frac{\sigma}{\sigma_Y} \right) - 1 \right]$$

The model was further analysed by Burdekin and Stone[15] to give the crack-tip opening-displacement as

$$\delta = \frac{8}{\pi} \frac{\sigma_Y}{E} a \ln \sec\left(\frac{\pi}{2} \frac{\sigma}{\sigma_Y} \right)$$

Wells suggested (without reference to the Dugdale model initially) that this crack-tip opening displacement (COD) could be used to characterise the mechanical conditions associated with fracture, in that it provided a direct indication of separation distance in the critical region (a bit like elongation or reduction of area at fracture in the tensile test).

The COD characterisation has at least two advantages. Firstly, the model is not restricted to small plastic zone size, but is valid right up to the point where the applied stress reaches the yield strength. The only proviso is that the real plastic zone should look like the strip-yield zone used in the model. Secondly, the COD can be directly measured in a fracture test (in which case it is designated δ_c), whereas the fracture toughness K_c can only be inferred from the fracture load through a theoretical solution which tends to break down as the plastic zone grows in size.

In fact, although this may have seemed to be a completely different approach to the fracture problem, it relates very firmly to the stress intensity method. For example, the equation for COD can be expanded in series form to show that

$$\delta = \frac{K^2}{E\sigma_Y} \left[1 + \frac{\pi^2}{24} \left(\frac{\sigma}{\sigma_Y} \right)^2 + \frac{\pi^4}{360} \left(\frac{\sigma}{\sigma_Y} \right)^4 + \cdots \right]$$

where K is the linear stress-intensity factor. The second term in the series corresponds closely to the 'first-stage' plastic zone size correction factor (see Section 5.1.3) and the further terms could be thought of as further iterations.

Of course, ideally one wants to extend this approach to the various finite geometries of interest, in the manner of *Figures 5.7* to *5.9*, but although this is possible in theory, only a few solutions have been published in a convenient form. *Figure 5.19* gives three important results. For the Dugdale model equivalents of cases 1, 3 and 4, approximate closed-form representations can be generated through the following equation (Gray[16]):

$$\rho = a \left[\sec\left(\frac{\pi}{2} \frac{\sigma}{\sigma_Y} \right) - 1 \right] F\left(\frac{a'}{W} \right) S$$

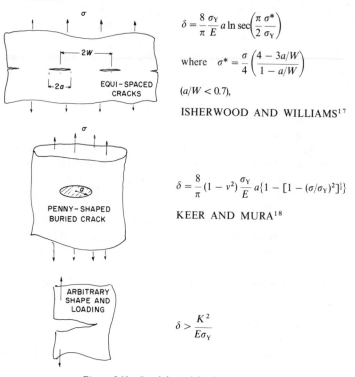

$$\delta = \frac{8}{\pi} \frac{\sigma_Y}{E} a \ln \sec\left(\frac{\pi}{2} \frac{\sigma^*}{\sigma_Y}\right)$$

where $\quad \sigma^* = \dfrac{\sigma}{4}\left(\dfrac{4 - 3a/W}{1 - a/W}\right)$

$(a/W < 0.7)$,

ISHERWOOD AND WILLIAMS[17]

$$\delta = \frac{8}{\pi}(1 - v^2)\frac{\sigma_Y}{E} a\{1 - [1 - (\sigma/\sigma_Y)^2]^{\frac{1}{2}}\}$$

KEER AND MURA[18]

$$\delta > \frac{K^2}{E\sigma_Y}$$

Figure 5.19   Dugdale model solutions

where

$$\frac{a'}{W} = \frac{a}{W}\left[1 + 0.2\left(\frac{\sigma}{\sigma_Y}\right)^P\right]\left[1 - \left(\frac{\sigma}{\sigma_Y}\right)^Q\right]^{-1}$$

and

$$\delta = \frac{8}{\pi}\frac{\sigma_Y}{E} a \ln \sec\left(\frac{\pi}{2}\frac{\sigma}{\sigma_Y}\right)F\left(\frac{a^*}{W}\right)$$

where

$$\frac{a^*}{W} = \frac{a + R\rho}{W}$$

Here, P, Q, R and S have different values for each geometry according to *Table 5.1*. The function $F(a'$ or $a^*/W)$ is exactly as quoted in *Figures 5.7–5.9*.

Table 5.1

Geometry	Limit load when	P	Q	R	S
			Parameters		
Centre crack	$\sigma/\sigma_Y \geqslant (1 - a/W)$	0.6	1.55	$\left.\begin{array}{c}0.2\\0.4\end{array}\right\}^a$	1.0
Double edge crack	$\sigma/\sigma_Y \geqslant (1 - a/W)$	5.1	1.68	0.5	1.0
Single edge crack	$(a + \rho)/W \geqslant 1^b$	0.6	1.6	0.47	0.82
Circumferential crack	$\sigma/\sigma_Y \geqslant (1 - a/W)^2$	0.17	1.41	0.36	0.85

^{a} Use $R = 0.2$ if $a/W \leqslant 0.8$ *or if* $a/W > 0.8$ *but* $(a + \rho)/W < 1$. Use $R = 0.4$ and set $(a + \rho)/W = 1$ if $a/W > 0.8$ and $(a + \rho)/W > 1$.
^{b} Or $\sigma/\sigma_Y \geqslant (1 - a/W)^2$.

Figure 5.20 shows the determination of crack-opening displacement in a finite-width edge-cracked plate through the previous equations. Notice that the limit load sets an upper boundary on the Dugdale model solutions (in the case shown limit load is reached when the net stress on the uncracked section reaches the yield strength).

Of course, even the strip-yield model in a sense fails to describe *general* yielding, that is, the point where the whole body is at or above the uniaxial yield strength in plane stress, and this is something of a limitation in relation to applications which involve large plastic strain. Notice, for example from *Figures 4.56* and *4.61*, that it is possible for regions of welded structures to experience plastic strains beyond yield without general collapse, especially in un-stress-relieved cases.

The strip-yield plastic zone size may also be used to index triaxial state in the manner previously described for LEFM. In *Figure 5.17* referred to earlier, the triaxial states at fracture, according to K_{Ic} or δ_c criteria, are mapped as a function of applied stress and crack size to thickness constraint ratio for an infinite plate. *Figure 5.21* shows the same kind of map for a finite-width internally-cracked geometry.

5.1.6 The *J*-integral generalised approach

Considerable progress had been made towards the description of crack-tip stress states by the mid-nineteen-sixties. It was however of some concern that these descriptions were least satisfactory in the region of most interest near the crack tip, in that they gave no picture of the stress state in the plastic zone. However in 1967, Rice[19]

244

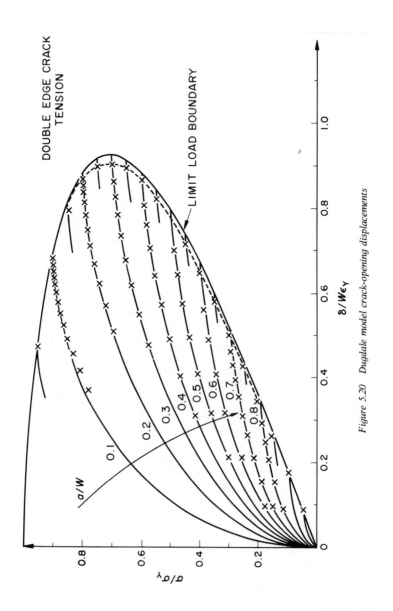

Figure 5.20 Dugdale model crack-opening displacements

245

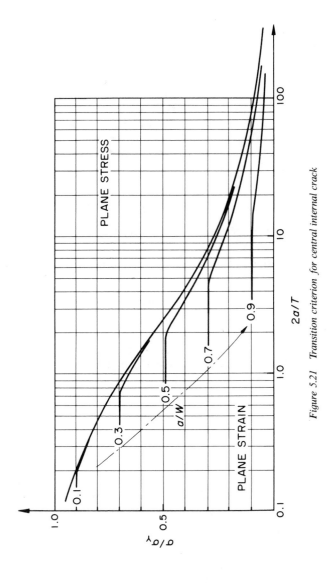

Figure 5.21   Transition criterion for central internal crack

formulated an important computational theorem which greatly expanded the range of methods which could be applied to crack problems, and also unified the previous approaches to characterisation. He showed that a certain combination of strain and potential energy quantities, when integrated along paths surrounding the crack tip, was 'path-independent', that is, the same numerical result was obtained whether the path was chosen through the near-crack-tip material or through material well away from the tip. This statement holds true whether the material is linear or non-linear. The integral was designated 'J' (units N/m). Several techniques for the calculation of J have since been advanced, including integration along paths drawn through finite-element grids (see Section 4.1.10), but the following conclusions are important.

If a near-crack-tip path is chosen in a linear elastic solution, the result

$$J = K^2/E$$

is obtained. Clearly K could be modified to include plastic zone size corrections.

Integrating along the top and bottom edges of the strip-yield zone in a Dugdale model gives

$$J = \sigma_Y \delta$$

a conclusion which also supports the earlier stated correspondence between K_p and δ.

J can also be determined experimentally, even for a non-linear material, by measuring the difference in work input to separate cracked bodies which have slightly different crack lengths (see *Figure 5.22*). Taking this idea to its extreme, if the limit load P_L on the cracked body varies with crack size

$$J = dP_L/da$$

Of course it may be noticed that the J-integral approach as such adds nothing to understanding of the crack-tip stress state. However in 1968 two solutions were produced for non-linear work-hardening material following the law $\varepsilon = B\sigma^n$ (one of them using the J-integral concept: references 20 and 21), and it became evident that identical near-crack-tip states do exist in such materials, given by a description of stresses,

$$\sigma_{x,y,xy,\text{etc}} \simeq \left(\frac{J}{B}\right)^{1/(n+1)} \frac{1}{(2\pi r)^{1/(n+1)}} f(\theta)$$

The existence of this unique stress pattern does much to explain the

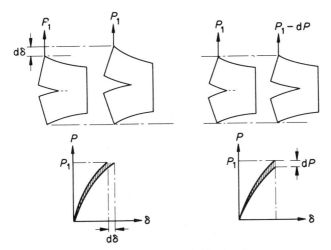

Figure 5.22 'Compliance' definition of *J*

apparent success of earlier approaches despite academic objections. Like the linear elastic *K* characterisation, problems reduce to determination of *J* which is a function at least of load, geometry and crack size as before but can now be extended to include material constitutive properties such as yield strength and work-hardening.

The yielding-strip model solutions are limited to components where the applied boundary stresses are less than the material yield strength. However, Burdekin and Stone have re-expressed the Dugdale solution to show the relationship between applied *strain*—measured over some appropriate gauge length—and crack-opening displacement. With some assistance from experimental evidence this formulation provides a useful guide to crack-opening displacement in regions of high stress concentration where the ratio $\varepsilon/\varepsilon_Y$ is often controlled by the SCF (see Chapter 4). Later results in the form known as the 'COD design curve' are shown here in *Figure 5.23*.

The crack-tip triaxial state is even more difficult to describe and predict in the general yielding case. Although triaxiality is probably not so severe as in the plane strain–small scale yielding examples, toughness transitions and interactions with component thickness still occur at general yield. Wells[22] has suggested that, in such cases, a nominal plastic zone size to thickness ratio of less than unity still forms a practical criterion to distinguish severe restraint.

248

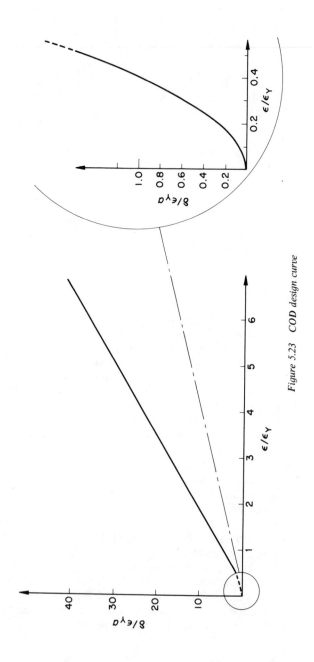

Figure 5.23 COD design curve

5.1.7 Summary

The crack-tip stress and strain parameters K, δ and J all provide a measure of the conditions which tend to stimulate crack propagation. They allow comparison of real components and convenient test specimens. The parameter K is appropriate to fractures which occur at gross stresses which are a small fraction of the material yield strength. The effect of crack-tip plasticity is to increase the effective crack length by a factor $(K^2/2\pi\sigma_Y^2)$.

The hydrostatic stress state associated with plane-strain constraint, gives way to a less onerous triaxial state (plane stress) when the nominal plastic zone size approaches the plate thickness:

$$(K/\sigma_Y)^2 \to \pi T$$

The unique stress state associated with linear elastic materials and the parameter K, breaks down near general yield in elastic/perfectly plastic materials. The crack-tip displacement δ—or the energy parameter J—provide bases for comparison in such cases.

5.2 Application to brittle fracture

The possibility that a welded structure may fail by brittle fracture is one which the designer must take very seriously. Typical failures have been spectacular and dangerous to life. The reader who is unfamiliar with the phenomenon ought to read some of the well-known case histories (references 23–29) and take any opportunity to examine a casualty structure at first hand.

The principal characteristic of brittle fracture is instability. An existing crack suddenly propagates through the material at a velocity which may be of the order of the speed of sound in the material. The adjective 'brittle' is used in a relative sense to suggest that there is little evidence of plastic flow or ductility (necking or tearing) before or during the fracture. It is, of course, possible for 'ductile' fractures to be unstable but as these will occur at loads which are nearer the general yield or limit load, they are less remarkable.

Brittle unstable fractures are not necessarily associated with shock loading, low temperatures, or obviously brittle materials such as glass; indeed in many of the unexpected failures, these features were absent.

5.2.1 Fracture toughness

The application of fracture mechanics to brittle fracture control is clear enough in principle—the level of crack-tip parameter existing at a known or postulated crack is compared with a fracture toughness number obtained from a test of the material to be used. However, toughness, like yield strength, is not a fixed quantity, but can depend on metallurgical variations, or mechanical and environmental factors.

It was shown in Chapter 2 that metallurgical structure varies greatly across a fusion weld from centreline to heat-affected zone. Add to this the fact that the various microstructures will be anisotropic, and one might imagine that a costly and laborious programme of toughness tests may have to be instituted to assess just one material and welding process. In practice, however, it is usually possible to make a preliminary survey of the weld section, and identify the most susceptible regions (probably those showing high hardness, coarse grain or low impact energy absorption in a Charpy test—see later). In steel welds, experience has shown that the heat-affected zone is particularly suspect, particularly when the effects of cyclic strain ageing (see p. 215) have not been removed by post-weld heat treatment. This is also a likely location for cracks, either parallel or perpendicular to the weld line. A familiarity with welding procedures and typical defects is useful, as an intelligent anticipation of defect locations can then be made.

In some metals, notably those showing a body-centred cubic crystal structure such as carbon steel, toughness is highly sensitive to temperature and strain rate. (At the microstructural level, dislocation movement seems to be inhibited at low temperatures or if the loading rate is high. Hence shear flow is limited and ductility suffers.) A small drop in temperature, or increase in loading rate, can have a drastic effect on toughness. Even when the loading rate is slow, the strain rate at the tip of a spreading crack will be much greater than when the crack is stationary. Hence toughness drops as the crack accelerates and a catastrophic running fracture may develop. In the extreme case where the load does not drop as the crack extends, crack arrest cannot take place. This can cause fractures several miles long in gas pipelines, where the running crack outstrips the depressurising wave, so that the crack tip always encounters pressurised pipe.

Temperature and strain rate effects are much less marked in face-centred cubic metals; for example, aluminium, copper and austenitic steel. For this reason these metals are often specified for low-temperature use. Large toughness variations can still be found in such materials, however, (see *Figure 5.15*) on account of triaxial transitions.

Many other specialised environments can alter toughness; for example, irradiation in nuclear containment vessels, chemical attack, or absorption of gases.

Before a given toughness number can be confidently used in fracture control, therefore, it must be carefully assessed as to its relevance to the specific fracture problem. The material tested must be metallurgically similar to the real structure, it must be subject to the same environment, and the crack-tip triaxiality must be reproduced in the test specimen. This last condition may be more difficult to satisfy than it seems. It is all too easy for plane strain to be maintained to high loads in a thick massive component which provides the greatest possible constraint on the plastic zone. It is quite another matter to reproduce this state in a laboratory specimen which must be of a practical finite size if it is not to be an embarrassment. If, as a result plane-strain constraint relaxes earlier in the test specimen than in the structure, a dangerously optimistic toughness will be indicated. As a general rule, therefore, the test specimen thickness should match the real structure. For similar reasons, fatigue loading is often used to prepare the test specimen, so that a realistically sharp initiating crack is incorporated.

5.2.2 Toughness testing procedures

Assuming that the conditions previously discussed are met, the conduct of a toughness test is simple enough in principle. The toughness specimen is subjected to steadily increasing load until crack propagation occurs and the level of crack-tip stress-intensity, opening displacement, or J value associated with this event is defined to be the *critical* value for that material in that environment. Most procedures make use of an autographic load–displacement record of the test (see *Figure 5.24*). In practice it is often difficult to determine what is actually happening at the crack tip and a wide range of behaviour has been observed in different materials and testing situations. At one extreme in very brittle materials one can describe events in terms of a simple picture whereby the behaviour is strictly linear until unstable fracture suddenly intervenes. At the opposite extreme, the crack grows slowly for a distance, while the load and hence the effective toughness increases steadily. This raises problems for the user—which point should be designated 'critical'? The level of K when the crack just starts to increase in length (assuming this can be identified) or the value when the behaviour goes unstable? Confidence in the latter possibility diminishes when one learns that the point of instability depends as much on the conditions of the test,

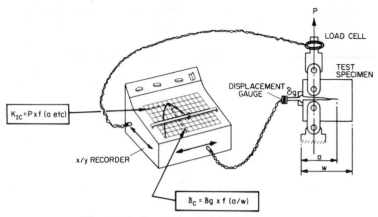

Figure 5.24   Autographic load-displacement record

for example, the initial crack size, the specimen thickness, loading configuration and testing machine stiffness, as it does on the 'intrinsic' material response. In fact, while these considerations raise very real problems which are still being considered and debated, for most practical purposes it is possible to classify the different kinds of behaviour into a limited number of categories which are in a sense idealisations of more complicated events.

The simplest kind of behaviour to consider approximates to the simple linear elastic instantaneous fracture described earlier. In metals and welded joints, this is most likely to occur in thick sections when the yield strength is high and the toughness is low, so that a linear elastic plane-strain mechanical state pertains. If the testing machine maintains the load as the crack starts to propagate and the testing configuration is one where K increases with crack length (i.e. *not* as given in *Figure 5.6b*), the strong likelihood is that the crack length at instability will be insignificantly greater than the initial crack length, especially if the material is rate-sensitive so that toughness drops as the crack propagates. Such a fracture is called a 'linear elastic' or 'plane-strain' fracture. The toughness is calculated from the load at instability and the initial crack length, and is designated 'K_{Ic}' because the fracture surface is usually predominantly mode I (i.e. no shear).

Of course it is important to establish precise limits on linearity, thickness, etc. which define whether or not the test result is 'valid' according to the categorisation described. Procedures have been developed (Srawley[30]) which place a limit on the plastic zone size and the degree of non-linearity to ensure that the crack cannot have

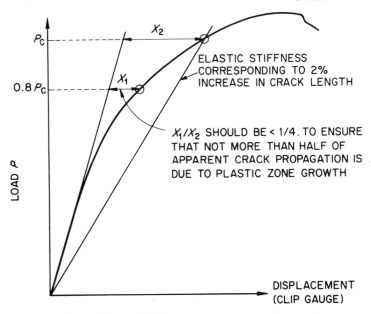

Figure 5.25   ASTM offset procedure for plane-strain tests

grown significantly before instability. There is also a severe limit on the plastic zone size to thickness ratio at fracture ($K_{Ic} < \sigma_Y (T/2.5)^{\frac{1}{2}}$) to make sure that plane-strain conditions were in force at the point where fracture is supposed to have occurred. Hence if a clear instability is identified on the test record, by a sudden load drop for example, the toughness can be calculated from this load and crack length, and admitted as a valid K_{Ic} result if it satisfies the restrictions on non-linearity, plastic zone size and plane-strain constraint (see *Figure 5.25*).

If a clear instability cannot be identified within the conditions given, but the fracture nevertheless corresponds to LEFM conditions as judged by plastic zone size, the next possibility is to accept that it may be more realistic to measure toughness as a variable quantity in terms of a 'crack growth–resistance curve'—an idea which was originated in relation to the testing of relatively thin sheet (Krafft *et al.*[31]) often under plane-stress conditions. In this approach, the loads and corresponding crack length are monitored continuously as the crack grows, and the value of K (or equivalent) characterisation parameter at each stage is plotted as in *Figure 5.26a*. If all goes well, the resistance curve should be independent of initial crack size as shown. However, the *point of instability* is not independent of initial

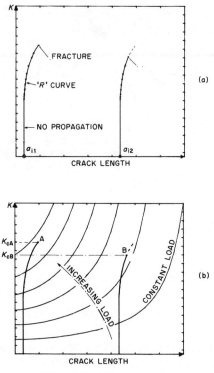

Figure 5.26   R-curve procedures

crack size according to the following explanation; consider two specimens which have different initial crack lengths but have the same crack growth resistance characteristics as shown in *Figure 5.26b*. If we superimpose on these *R*-curves a family of stress-intensity curves which show how applied *K* increases with crack length for given constant loads, it can be seen that for all stages up to points A and B the crack growth resistance increases faster than the stress intensity. Therefore growth should be stable up to these points. Above A and B, crack growth will be unstable unless the load drops. Therefore different family of stress-intensity curves could give yet another value K_{cB}), and a change in specimen configuration which produces a different family of stress intensity curves could give yet another value of K_c. Despite these complications, however, it also has to be said that for many low-toughness materials tested in wide-sheet form, the instability point is not very sensitive to initial crack size and in such cases it may be quoted as K_c or 'plane-stress toughness'. (The subscript I is dropped, as the fracture mode may not be flat.)

If the behaviour of the toughness test cannot be fitted into either of the preceding schemes, the implication is that the material and configuration are sufficiently ductile to survive into the regime where elastic/plastic or general-yielding fracture mechanics is required to describe events and universally accepted procedures are not so easily found. The first possibility is to measure toughness in terms of J_c. In this case, J at the point of instability may be inferred from the load via an appropriate solution or from an experimental calibration. Again similar problems arise to the K_{Ic}/R form of testing. How does one identify crack propagation and in this case how can non-linearity due to crack growth be distinguished from non-linearity due to growth of the plastic zone at a stationary crack tip? Sumpter and Turner[32] discuss some of these problems and their solutions.

The crack-opening displacement concept on the other hand skirts many of these imponderables because the displacement can be measured directly as distinct from other procedures which infer values from the current load. In practice, it is difficult to measure this displacement at the true crack tip in a routine test because the initial crack is too fine and therefore procedures have been developed in which the opening is measured at a wider part of the starter-crack (the crack-mouth) and the crack-tip opening is inferred through geometric calibrations (reference 33). This procedure, like many of the J_c methods which employ load-point displacement, creates difficulties for the observer in that measurements made a long way from the crack tip are relatively insensitive to events such as triaxial transition and slow crack growth. Therefore an alternative procedure has been suggested in which the thickness or 'notch-root contraction' is measured at the true crack tip using a simple transducer (Gray[34]). This measurement turns out to be numerically equal to COD and therefore can be used to give a sensitive indication of crack-opening displacement at the point of first separation. In both the J_c and δ_c approaches the occurrence of slow crack growth leads to the same difficulties in terms of interpretation and application as found in LEFM—that is, should one use the value when the crack first starts to propagate or 'initiate' (δ_i, J_i) or is it reasonable to use the value at instability, recognising that this may not be an invariant property? No universally accepted understanding of these points has emerged so far, but both concepts are widely used as simple toughness numbers without too much anxiety as to the finer points.

5.2.3 Qualitative testing

There are some toughness tests and associated fracture control philosophies which predate the strictly numerical approach of sharp-crack mechanics. They represent various attempts to reproduce known features of brittle fracture in a relatively controlled and scientific manner.

The Charpy impact test became very popular in association with a 'transition temperature' philosophy after many post-war investigations into the welded Liberty ship accidents. In this test a falling pendulum hammer is used to fracture a 10 mm square section three-point bend specimen, which has a machined notch of 2 mm tip radius. The energy absorbed in the fracture process (plus friction losses) is measured and a typical test result for a carbon steel is shown in *Figure 5.27*. A rapid increase in toughness is seen over a small

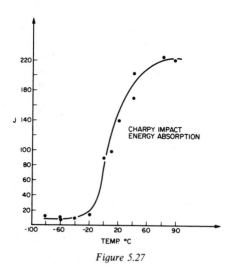

Figure 5.27

increase in temperature. In its straightforward form, the transition temperature philosphy requires that a given steel should not be applied in a service environment below its transition temperature, regardless of crack sizes, applied stresses or loading rates. There are obvious objections to this simple idea however, the most serious being that the 10 mm square specimen is too small to maintain plane-strain conditions relative to thick structures. (Another way of expressing this is that thicker specimens would demonstrate higher transition temperatures.) The high loading rate may also be unrealistic for slowly loaded structures, although it obviously errs on the safe side. The notch is also insufficiently sharp. It is obviously sensible to operate the material on the 'upper shelf' of the transition curve (however defined) but there is no guarantee that the corresponding toughness is sufficient to resist fracture in the presence of a given crack size or applied stress.

Nevertheless, the Charpy impact test does rank material toughness in the correct order, and it can, therefore, be usefully applied if the test results can be related to other full thickness tests or to measured service experience. Such an approach is adopted in BS 4642 and described in Wells[11]. It is attractive as a quality control test, and its use in sampling weld zones has already been noted.

Wells and others have drawn the Charpy energy measurement into the general scheme of fracture mechanics, and although their arguments must be applied with caution, as there are restrictive assumptions, some interesting trends can be shown. In the plane-stress regime the critical crack-opening displacement δ_c is proportional to Charpy energy/yield strength. Therefore, if uniform resistance to fracture is desired, the specified energy absorption should be increased with material yield strength. If sufficient toughness is required to maintain the structure above the plane strain–stress transition point or to satisfy similar criteria in terms of critical crack length to thickness ratio, the energy requirement is even more exacting—the proportionality above can be combined with the approximate equation $\delta_c = K_c^2/E\sigma_Y$ and the transitional plastic zone size to thickness ratio $(K/\sigma_Y)^2 > \pi T$ to demonstrate that the Charpy energy required is proportional to $\sigma_Y^2 T$. Correlations for a range of pressure vessel steels are given in an interesting figure in Wells[11]. It is perhaps significant that recent pressure vessel fractures have been associated with thick, high yield strength steels.

5.2.4 Residual stress

In many structures, it is considered impractical or too costly to apply a residual-stress-relieving treatment (thermal or mechanical) before the structure enters service. Experience shows that such treatments greatly reduce the risk of brittle fracture, especially in components thicker than 20 mm. It is obvious enough that a procedure which reduces tensile stresses in potentially cracked locations will be beneficial, but in the case of thermal stress-relieving, there are also metallurgical reasons; if the structure is uniformly heated to a temperature which allows significant creep (greater than 600 °C for steels), then much of the metallurgical damage which may be caused by repeated plastic straining and ageing is repaired. Thus, a substantial improvement in fracture toughness may be obtained incidentally.

Nevertheless, an approach to fracture control is needed for structures which have not had the benefit of stress relief. If the residual stress pattern transverse in the uncracked plate is known and a crack is then assumed to be introduced transverse to this stress, the effect may be modelled by reversing the direction of the residual stress on the crack edges. This produces the correct stress-free surface boundary condition. The linear mechanics solution for pairs of opposed 'prising-apart' forces is then integrated to give the total stress intensity due to residual stress. The results of such a calculation are shown in *Figure 5.28*. It is interesting to note that the combination of

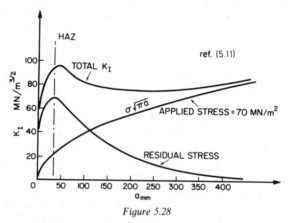

Figure 5.28

high stress intensity and poor toughness in the heat-affected zone could well lead to spontaneous fracture. If additional service loading is now applied (70 MN/m^2) in the above example, K_I may be found by superposing solutions as described on p. 230.

An alternative method of attack (Burdekin and Dawes[35]), assumes quite reasonably that a problem crack could be expected to be entirely embedded in a region which is sustained at general yield. Thus, for example, consider a loading which would normally generate an applied strain normal to the crack of three times yield strain: if the component is to be assessed in the as-welded condition, then it would be treated with the aid of *Figure 5.23* as if a strain of four times yield existed.

5.2.5 Wide-plate tests

It is probably clear that the fracture assessment of as-welded structures will usually be frustrated by insufficiently detailed input data (residual stress pattern, welding damage, etc.). In such cases, the results of wide-plate fracture tests are especially valuable.

In Britain, the form of test developed by Wells at the Welding Institute is the most familiar. A large full-thickness plate, typically one metre square, is butt welded, using the welding process and treatments which are to be assessed. Notches are cut in the plate edges before welding, and transverse to the weld direction. The notch tips suffer the plastic straining and thermal cycling which accompany fusion welding processes, and reproduce in an arbitrary way the possibility of cracking at an early stage of fabrication. The completed plates are stress-relieved or not as the case may be, and placed in a massive tensile testing machine, to be tested at a representative temperature. The test results are usually quoted in terms of stress or plastic strain at fracture.

The benefits of stress relief are illustrated in *Figure 5.29* which is calculated from wide-plate test data given in Burdekin and Wells[36]. Critical crack-opening displacements are derived from the reported crack sizes and measured strains, via *Figures 5.7–5.9* (for equivalent crack size) and *5.23*. The effect of the recommended adjustment for residual stress (described three paragraphs previously) is also shown.

The deterioration in fracture toughness can be simulated for the purpose of a small-scale test by prestraining and ageing a suitably notched specimen. The broken line in *Figure 5.29* was obtained from Saunders and Dolby[37] and demonstrates that such a simulation can provide useful information.

5.2.6 Good practice in fracture safe design

The numerical examples which conclude this chapter may distract attention from some of the simple conclusions which can be drawn from the previous discussion, and which could form useful guiding principles:

1 Applied stress and crack size are the important parameters in brittle fracture. Consider whether the welds cannot be placed in alternative low-stress regions, or whether residual welding stresses cannot be reduced by mechanical or thermal stress relief.

2 Take care that the metallurgical quality associated with the material and the proposed welding procedure and any subsequent heat treatment is firmly established and controlled in production. Some recent service fractures have been stimulated by unexpected zones of poor weld toughness. In this context one should realise that a material which has poor weldability or

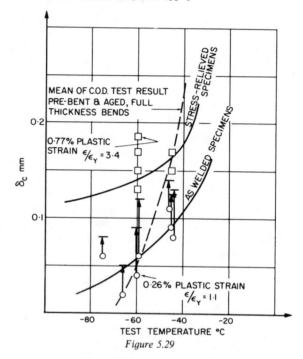

WIDE PLATE TESTS ON
55-75 mm M$_n$ -C$_r$ -M$_O$-V
LOW ALLOY STEEL, WELDED BY
M.M.A. PROCESS.

☐ STRESS-RELIEVED AT 550° OR 600°C

┬ ADJUSTMENT FOR RESIDUAL STRESS

◯ AS-WELDED OR S.R. AT 450°C

Figure 5.29

requires special fabrication controls, may be a false economy.

3 Remember that the toughness required to tolerate a given stress
and crack size increases in proportion to yield strength and
thickness. Extrapolation to higher yield strength material, higher
design stresses and greater thickness must be made warily with
these trends in mind.

Example 5.1 Linear fracture mechanics

A tank for the storage of liquefied gas at low temperature is made
from 40 mm thick aluminium alloy (yield strength 240 MN/m^2) and

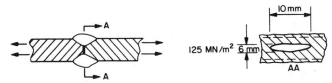

Figure 5.30

incorporates a double-sided gas–metal arc weld as shown in *Figure 5.30*. An equipment malfunction results in a typical lack-of-interpenetration defect as shown. Estimate K_I for the defect illustrated. If a toughness test is to be made to measure K_{Ic}, prescribe the important test conditions.

Answer:

Note that the shorter axis of a planar defect is the more significant dimension (see case 10). Using the infinite-plate solution to obtain a first estimate:

$$K_I = \sigma(\pi a)^{1/2} = 125(\pi \times 3 \times 10^{-3})^{1/2} = 12.1 \text{ MN/m}^{3/2}$$

Applying corrections for the elliptical shape and the finite geometry (that is to account for the nearness of the crack tips to the plate surface)

$$K_I = \sigma[\pi a F_1(a/W)F_{10}(a/b)]^{1/2} = \sigma[\pi a F_1(\tfrac{3}{20})F_{10}(\tfrac{3}{5})]^{1/2}$$

$$= 125[\pi \times 3 \times 10^{-3} \times 1.04 \times 0.629]^{1/2}$$

$$= 9.81 \text{ MN/m}^{3/2}$$

An approximate correction for plasticity can be included by replacing *a* where it appears in the equation above by $a^* = (a + K_I^2/2\pi\sigma_Y^2)$ giving

$$K_I = \sigma[\pi a^* F_1(a^*/W)F_{10}(a/b)]^{1/2}$$

$$= 10.25 \text{ MN/m}^{3/2}$$

Note that r_Y is much smaller than *a* in this case, which validates the correction procedure. It is unnecessary to 'correct' the ratio a/b.

The three-dimensional state of the crack tip can be indexed by examining the ratio of the plastic zone size to the distance to the nearest free surface (in this case the welded ligament). The plastic zone size is 0.26 mm which is much less than the arbitrary limit of $0.4 \times 17/2\pi$ quoted in Section 5.1.4 for the achievement of plane-strain conditions. Also, given a σ/σ_Y ratio of 0.52 and a $2a$/ligament ratio of

0.35, *Figure 5.21* tends to confirm the diagnosis of a plane-strain mechanical state.

Therefore to give a result which is relevant to the service situation:

1 The toughness specimen should demonstrate plane-strain constraint at least up to 10.25 MN/m$^{3/2}$.
2 The test should be carried out at the service temperature.
3 The specimen should be notched in weld metal along the weld centreline.

Example 5.2 Transitions

The HAZ of a weld in a given material has a yield strength of 400 MN/m^2 measured at 0.2% permanent offset. Toughness tests are carried out in which the material survives to a K_I level of 60 MN/m$^{3/2}$ in plane strain and on further increase of loading a transition occurs. Eventually the crack starts to propagate at an indicated δ_c of 0.3 mm.

Estimate critical crack sizes for 'infinite-plate' structures incorporating such welds, where

(a) the structure is 100 mm thick and the applied stress is 0.3 of the yield strength;
(b) the thickness is 10 mm and the load is increased to 0.7 of the yield strength.

Answer:

For the first structure

$$K_{Ic}^2/\sigma_Y^2 T = (60/400)^2/0.1 = 0.225$$

and

$$\delta_c/\varepsilon_Y T = 0.3/0.002/100 = 1.5$$

Hence from *Figure 5.17*, plane-strain fracture could occur at $\sigma/\sigma_Y = 0.3$ for a crack length of 160 mm ($2a/T \simeq 1.6$). For greater crack lengths, plane-strain fracture could intervene before the situation is relieved by a triaxiality transition.

For the second structure,

$$K_{Ic}^2/\sigma_Y^2 T = (60/400)^2/0.01 = 2.25$$

and

$$\delta_c/\varepsilon_Y T = 0.3/0.002/10 = 15$$

The corresponding critical crack length in plane stress is 150 mm and the structure should reach this point as a transition will occur at $\sigma/\sigma_Y \simeq 0.28$ with $K^2/\sigma_Y^2 T < 2.25$.

Example 5.3 General yielding: residual stress assessment

The fracture toughness of a 25 mm thick low-alloy steel (yield strength 460 MN/m^2) is sampled at various stages of a fabrication procedure. Each sample is in the form of a 10 mm square bend specimen, with a 3 mm notch located in the HAZ of the weld. Tests conducted at $-20\,°C$ give the following results:

Stage	δ_c mm (average)
After welding	0.18
500 °C heat treatment	0.23
650 °C heat treatment	0.5

Welds are specified for locations where the applied strain is controlled by a geometrical feature to twice yield. Estimate critical sizes for the various stages of fabrication and for operational conditions.

Answer:

The first important consideration in this case is that the test specimen is much thinner than the real structure and hence plane-strain constraint will not be reproduced properly in the test to as high a level of toughness parameter. Using the arbitrary conservative limit quoted in Section 5.1.4, the maximum level of K_{Ic} which can be proved by the specimen is given by

$$K_{Ic} = \sigma_Y (0.4 \times \text{ligament})^{1/2} = 24.3\ \text{MN/m}^{3/2}$$

whereas in the structure, plane strain may persist to 46 MN/m$^{3/2}$ or even higher.

However, assuming for the present that the true K_{Ic} is high enough to permit safe transition to plane stress without fracture, the risk of spontaneous fracture during fabrication and before stress relief could be assessed in terms of an applied strain range $\varepsilon/\varepsilon_Y = 1$ and $\delta_c = 0.18$ mm, using the COD design curve (*Figure 5.23*), giving

$$a_{\text{crit}} = \frac{\delta_c E}{\sigma_Y \times 4.5} = \frac{0.18 \times 200 \times 10^3}{460 \times 4.5} = 17.4\ \text{mm}$$

Other possibilities are:

Stage	$\varepsilon/\varepsilon_Y$	δ_c (mm)	a_{crit} (mm)
Unstress-relieved operation	2 + 1 = 3	0.18	4.6
Operation after 500 °C treatment			
assuming no residual stress reduction	2 + 1 = 3	0.23	5.9
Operation following 650 °C treatment	2	0.5	19.8

Example 5.4 Specification of basic toughness

A 35 mm thick steel of 400 MN/m² yield strength is to be specified to reduce topside weight in offshore floating structures. The following points are raised.

In previous similar structures where steels of less than 25 mm thickness and 250 MN/m² yield strength were used, minimum toughness was arbitrarily specified in terms of a Charpy impact energy absorption (C_v) of 30 J at the service temperature. These structures showed a high but not unreasonable tolerance to large cracks. Such cracks commonly developed due to fatigue loading in locations which were not convenient for periodic inspection. How can previous experience be extrapolated?

Making use of Wells' correlation between Charpy impact energy absorption and $\delta_c(\simeq \frac{1}{2}C_v(\sigma_Y \times \text{ligament area}))$ the material formerly used had a non-dimensional toughness identified by:

$$\frac{\delta_c}{\varepsilon_Y T} = \frac{1}{2} \times 30\left[\frac{\sigma_Y^2}{E}\frac{10 \times 7}{10^6}T\right]^{-1}$$

$$= \frac{1}{2} \times 30\left[\frac{(250 \times 10^6)^2}{200 \times 10^9}\frac{10 \times 7}{10^6}\frac{25}{10^3}\right]^{-1}$$

$$= 27.4$$

This toughness level would place an incipient fracture well into the plane-stress regime (see *Figure 5.17*). A similar reserve with respect to the transition boundary would be exhibited by steel proposed for use if it had an impact energy absorption given by

$$30 \times \left(\frac{400}{500}\right)^2 \times \frac{35}{25} = 108 \text{ J}$$

This of course implies a tolerance to cracks which are bigger than the existing situation by the factor $(108/30)/(400/250) = 2.25$.

REFERENCES

1. INGLIS, C. E., 'Stresses in a plate due to the presence of cracks and sharp corners', *Trans. Inst. Nav. Archs*, **55**, pt 1, 219 (1913)
2. GRIFFITH, A. A., 'The phenomena of rupture and flow in solids', *Phil. Trans. R. Soc., Lond.*, **A211**, 163 (1921)
3. IRWIN, G. R., 'Relation of stresses near a crack to the crack extension force', *Proc. 9th Int. Congr. on Appl. Mech., Brussels*, Paper 101(2), 245 (1956)
4. WILLIAMS, M. L., 'On the stress distribution at the base of a stationary crack', *J. Appl. Mech.*, **24**, 109 (1957)

5. PARIS, P. C., and SIH, G. C., 'Stress analysis of cracks', *ASTM Spec. Tech. Publ.* 381 (1965)
6. SIH, G. C., *Handbook of Stress Intensity Factors for Researchers and Engineers*, Inst. Fracture and Solid Mech., LeHigh University, Bethelehem (1973)
7. ROOKE, D. P., and CARTWRIGHT, D. J., *Compendium of Stress Intensity Factors*, HMSO (1976)
8. TADA, H., PARIS, P. C., and IRWIN, G. R., *Stress Analysis of Cracks Handbook*, Del Research Corp., Hellertown, Pa (1973)
9. TADA, H., PARIS, P. C., and IRWIN, G. R., *Stress Analysis of Cracks Handbook, Suppl.* 1, Del Research Corp., Hellertown, Pa (1973)
10. 'The compounding method of estimating stress intensity factors in complex configurations', Item 78036, Engineering Sciences Data Unit
11. WELLS, A. A., 'Notched bar tests, fracture mechanics and the brittle strengths of welded structures', *Br. Welding J.*, **12**, 1 (1965)
12. McLINTOCK, F. A., and IRWIN, G. R., 'Plasticity aspects of fracture mechanics', *ASTM Spec. Tech. Publ.* 381, 84 (1965)
13. IRWIN, G. R., 'Fracture mode transition for a crack traversing a plate', *J. Basic Engng*, **417** (1960)
14. DUGDALE, D. S., 'The yielding of steel sheets containing slits', *J. Mech. Phys. Solids*, **8**, 100 (1960)
15. BURDEKIN, F. M., and STONE, D. E. W., 'The crack opening displacement approach to fracture mechanics in yielding materials', *J. Strain Anal.*, **1**, 2 (1966)
16. GRAY, T. G. F., 'A closed-form approach to the assessment of practical crack propagation problems', *Fracture Mechanics in Engineering Practice*, Applied Science Publishers (1977)
17. ISHERWOOD, D. P., and WILLIAMS, J. G., 'The effect of stress–strain property on notched tensile failure in plane stress', *Engng Frac. Mech.*, **2**, 9 (1970)
18. KEER, L. M., and MURA, T., 'Stationary crack and continuous distribution of dislocations', *Proc. 1st Int. Conf. on Fracture, Sendai*, Vol. 1, p. 99
19. RICE, J. R., 'A path-independent integral and the approximate analysis of strain concentration by notches and cracks', *J. Appl. Mech.*, **35**, 379–86 (1968)
20. RICE, J. R., and ROSENGREN, G. F., 'Plane strain deformation near a crack tip in a power-law hardening material', *J. Mech. Phys. Solids*, **16**, 1–12 (1968)
21. HUTCHINSON, J. W., 'Singular behaviour at the end of a tensile crack in a hardening material', *J. Mech. Phys. Solids*, **16**, 13–31 (1968)
22. WELLS, A. A., 'Fracture control of thick steels for pressure vessels', *Br. Welding J.*, **15**, 5 (1968)
23. PARKER, E. R., *Brittle Behaviour of Engineering Structures*, Wiley (1957)
24. *BWRA Bulletin*, **7**(6) (1966)
25. TIPPER, C. F., *The Brittle Fracture Story*, Cambridge University Press (1962)
26. PELLINI, W. S., and PUZAK, P. P., 'Fracture analysis diagram procedures for fracture-safe engineering design of steel structures', *Welding Res. Council Bull.* 88, May (1963)
27. MOISIO, T., 'Brittle fracture in failed ammonia plant', *Metal Constr. Br. Welding J.*, **4**, No. 1 (1972)
28. COTTELL, G. A., 'Explosion of a large air-receiver of solid forged construction', *Br. Engine Tech. Rep.*, Vol. X (1971)
29. 'Report on the brittle fracture of a high-pressure boiler drum at Cockenzie Power Station', South of Scotland Electricity Board, Edinburgh (1967)
30. SRAWLEY, J. E., 'Plane strain fracture toughness', *Fracture*, Vol. IV, Academic Press (1969)

31. KRAFFT, J. M., SULLIVAN, A. M., and BOYLE, R. W., 'Effect of dimensions on fast fracture instability of notched sheets', *Proc. Crack Propagation Symp. College of Aeronautics, Cranfield*, Vol. 1, pp. 8–28 (1961)

32. SUMPTER, J. D. G., and TURNER, C. E., 'Method for laboratory determination of J_c', *Cracks and Fracture ASTM Special Tech. Publ.* 601 (1975)

33. 'Methods for crack-opening displacement (COD) testing', DD19, British Standards Institution (1972)

34. GRAY, T. G. F., 'Notch root contraction as a fracture measurement', *Int. J. Frac. Mech.*, **8**, 277–85 (1972)

35. BURDEKIN, F. M., and DAWES, M. G., 'Practical use of linear elastic and yielding fracture mechanics with particular reference to pressure vessels', *Inst. Mech. Engrs Conf. on Practical Applications of Fracture Mechanics to Pressure Vessels* (1971)

36. BURDEKIN, F. M., and WELLS, A. A., 'Wide plate tests on a Mn–Cr–MoV steel', *Br. Welding J.*, **13**, 2 (1966)

37. SAUNDERS, G. G., and DOLBY, R. E., 'Effects of welding and post-weld heat treatment on the fracture toughness of Mn–Cr–MoV steels', *Br. Welding J.*, **15**(5), 230 (1968)

38. GRAY, T. G. F., 'Convenient closed form stress intensity factors for common crack configurations', *Int. J. Frac.*, **13**, 65–75 (1977)

Chapter 6

Fatigue

Introduction

A component or structure which survives a single application of load, may fracture if the application is repeated a large number of times. This so-called fatigue failure seems to be an engineering blind spot, for it is the most common mode of failure in service. The presence of repeated loading provides the warning clue; bridges, cranes, ship structures, automotive and aero components, pressure vessels, all experience repeated loading, use welded construction and have a high fatigue failure rate. Fillet welded construction is particularly troublesome. Fatigue, followed by brittle fracture, can be a catastrophic combination, as exemplified by the Comet disasters of the 1950s.

Even when the danger has been fully recognised at the design stage, failure has occurred. This suggests that designing against fatigue is far from straightforward and probably requires more engineering experience and judgement than is often available.

Although a vast literature on fatigue has built up since Wöhler's famous work on railway axles in 1860, the engineer may find much of it confusing, and conflicting. It may be helpful to describe and discuss some of the fatigue phenomena relevant to welded construction before embarking on a practical design approach.

6.1 Understanding fatigue research

6.1.1 Phenomena

Fatigue test results are traditionally presented in the form of a Wöhler or S/N curve. The stress S is plotted against the number of load cycles N required to produce failure. For some materials, there exists a 'fatigue limit stress' below which fatigue failure will not occur.

This presentation is simple and useful, but may in itself be responsible for much of the confusion; the inference may be drawn

267

that a single stress-controlled failure mechanism persists throughout the life of the component, leading eventually to fracture. In fact, several mechanisms operate in sequence: initiation of a microscopic defect, slow incremental crack propagation, and final unstable fracture. The progress of each of these stages is influenced differently by the same forces and effects. Variations in each stage are easily

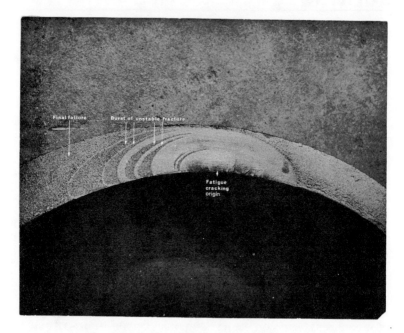

Figure 6.1   Stages of fatigue failure

concealed in the 'complete failure' S/N curve. In particular, where significant initial cracks are present (e.g. in welds) the initiation stage may be completely bypassed and the life thereby shortened. This kind of variation also accounts for much of the experimental scatter which is typical of fatigue experiments.

The various stages can sometimes be visually distinguished on the casualty fracture surface (*Figure 6.1*). The flat, regular and finely rippled or 'striated' surface provides a direct record of the successive positions of the crack tip during the incremental propagation stage. The initiating defect(s) can be recognised as the source of the rippled pattern. Where the specimen is initially crack-free, the initiating microcrack is generated on a maximum shear plane, and usually turns on to a tensile principal plane during subsequent propagation. There is often an obvious change in surface texture to allow distinction between the slow propagation stage, and the final unstable fracture which severs the component.

6.2 High-strain fatigue crack initiation

In the nature of things, welds will frequently be found at or near geometrical stress concentrations in the structure. They may also in themselves possess stress-concentrating features (e.g. the over-fill

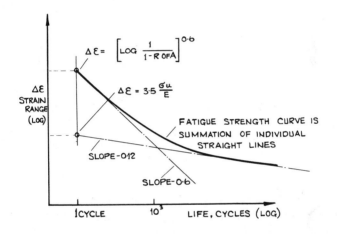

$$\Delta \varepsilon = \left[\text{LOG} \; \frac{1}{1 - R \; \text{OFA}} \right]^{0.6}$$

$$\Delta \varepsilon = 3.5 \frac{\sigma_u}{E}$$

FATIGUE STRENGTH CURVE IS SUMMATION OF INDIVIDUAL STRAIGHT LINES

SLOPE - 0.12

SLOPE - 0.6

Figure 6.2

shape). It is likely, therefore, that they will experience strains beyond yield, even when the bulk of the component is stressed to a low proportion of the material yield strength. High-strain fatigue cracks will be initiated very quickly at these points.

There has been a great deal of effort directed towards the prediction of high-strain fatigue strength from the more commonly quoted static tensile properties. A convenient summary of this work is given in reference 1. *Figure 6.2* (after Manson[1]) shows how a fatigue curve which predicts the life of small strain-cycled specimens can be constructed from two commonly quoted properties, reduction of area, and ultimate strength. Ductility is seen to be relevant to the low-life region (less than about 10^3 cycles) and ultimate strength (measured on the final area at fracture) to the higher-life region. It is interesting to use this empirical plot to compare a number of common steels as in *Figure 6.3*. Although the ultimate strengths covered range from 400 to 760 MN/m², the differences in fatigue performance are small. For medium-strength steels, the fatigue limit is found at about 2 to 5 million cycles (in other words a specimen reaching this life will probably survive indefinitely) and this corresponds to a strength of about 50 or 60% of the ultimate strength.

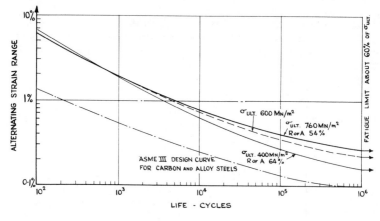

Figure 6.3

In the period since the first description of the Manson approach, it has become known that the 'static' tensile test should be conducted in a certain way to obtain good prediction[2]; there can be considerable differences between monotonic and cyclic stress–strain behaviour, and to reflect these differences the specimen should be cycled with increasing strain increments at a strain rate which compares with the rate to be used in the fatigue test. A summary of these methods is given in reference 3. Until data obtained in this special way is routinely available, however, engineers will probably have to make do with the more familiar static tensile properties, and accept some inaccuracies.

Yield strength is a familiar property which is significantly *absent* from the fatigue prediction. The obvious inference therefore is that fatigue strength in the high-strain region at least is unrelated to yield strength. Published evidence on the testing of full-scale structures[4,5] supports this view. It has also been recognised for some time that the incorporation of an unmachined weld reduces materials of widely differing yield strengths to a common level. Reference 6 attributes this levelling of strength to the strain-raising effect of small crack-like defects at the toes of welds.

Unfortunately, design stresses are usually increased when a higher-yield-strength material is introduced into an established design; indeed, this is the major economic justification for high-strength materials. If the working stress is thereby increased, the component will have an even shorter life. Of course, only the cyclic stress *range* is important, and a high-yield-strength material may be justified where a large component of steady loading exists.

6.3 Propagation (application of fracture mechanics)

Once a discernible crack is present (it may have been created by fatigue cycling or during some previous stage such as welding), it is important to know whether it will propagate further and, if so, how

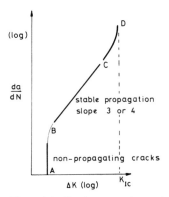

Figure 6.4   Crack propagation modes

quickly. One might guess that the answers to these questions will be related to the crack-tip stresses or strains, and in a number of hypothesised propagation mechanisms, the shear strains in the 45° plastic zones (see *Figure 5.16*) are considered to be significant[7]. In practice, plane-strain conditions will frequently apply and the linear crack tip stress intensity factor offers a convenient way of keeping track of, or describing, the crack-tip stresses. K has for some time now been used with advantage as a controlling parameter in fatigue crack propagation studies.

Experimental evidence at present suggests at least three distinct propagation modes corresponding to increasing levels of K. These are shown schematically in *Figure 6.4*, where propagation rate is plotted against stress intensity range.

The first portion of the curve AB shows that there is a minimum value of ΔK below which the crack will not propagate, regardless of the number of load cycles. Such a level has been reported by many authors; for example, reference 8 shows that an existing crack will not propagate if $\Delta K/E$ (a crack-tip strain intensity range) is less than 5×10^{-4} mm$^{\frac{1}{2}}$. This claim covers a variety of materials enclosing a fourfold range of ultimate strength and a fourteen-fold range of yield strength.

One may speculate that the crack-tip shear strains at the minimum level are insufficient to 'slide off' fresh material from the 45° zones on to the crack perimeter. The nonpropagating crack may also be linked with the fatigue limit strength in that the associated stress intensity is insufficient to propagate stage 1 cracks. In any case, this empirical finding allows a numerical expression of a principle which is well known in the motive power and aero industries, namely, that a component made of a high-strength alloy can only be operated at a correspondingly high stress if a high-quality surface finish is provided.

Example 6.1

A certain component is made of mild steel (yield strength 210 MN/m^2) and gives satisfactory service in a fatigue situation when stressed to 2/3 of its yield strength. It is replaced by a component of identical geometry made of a low-alloy steel with a 630 MN/m^2 yield strength, which is stressed to the same proportion of yield strength. Fatigue failures follow the change in design. Explain!

For both materials, the critical size of crack is found from

$$\frac{\Delta K}{E} = 5 \times 10^{-4}$$

i.e.

$$\frac{\Delta \sigma (\pi a)^{\frac{1}{2}}}{E} = 5 \times 10^{-4}$$

therefore

$$a_{\text{CRIT}} = \frac{25 \times 10^{-8} E^2}{(\frac{2}{3}\sigma_Y)^2 \pi}$$

as Young's Modulus will be equal for both steels,

$$a_{\text{CRIT MILD STEEL}} = 0.16 \text{ mm}$$
$$a_{\text{CRIT LOW ALLOY}} = 0.018 \text{ mm}$$

Whereas it might be feasible to fabricate and finish a component so as to exclude defects greater than 0.16 mm in length, a similar component in the low-alloy steel will have to be manufactured to exceptional standards (e.g. polish finish) to exclude defects greater than 0.018 mm.

When ΔK is increased from the threshold value, the crack lengthens during each load cycle. The propagation rate has been observed to be proportional to a power of ΔK (3 or 4) as shown in the region BC of *Figure 6.4*. This proportionality has been explained[9] in terms of the previously described model in which irreversible shear flow from the plastic zones adds to the crack perimeter.

For further increase of ΔK, as in regions CD, the propagation rate is proportional to higher powers of ΔK. Several mechanisms may be responsible, e.g. an element of stable 'mode I' cracking may be added to the previous 'sliding off' extension during each cycle, and materials which have a poor fast fracture toughness are particularly sensitive. Cracks propagate faster in these materials when the loading pattern contains a high proportion of tensile mean stress. Transition to plane-stress conditions with rising load, and strain rate effects may also play a part. Experimental data are given in references 9 to 16.

Eventually, the growing crack will reach a size at which unstable ductile or brittle fracture becomes possible. This point may be estimated from the plastic limit load, or from the appropriate fracture toughness. Neglecting the effect of finite specimen size (e.g. crack length to plate width ratio), the time spent in propagation will be the same for differing component thicknesses. It is therefore technically possible, and in many cases practicable, to predict the behaviour of a fatigue-loaded component from crack initiation through to the onset of unstable fracture. The following example is perhaps fanciful, but demonstrates the possibilities.

Example 6.2

The master of a ship arriving in a foreign port reports a crack in a 25 mm deck plate. The progress of the crack has been logged at weekly intervals over the last five weeks since discovery and the entries read: '234 mm, 245.6 mm, 257.6 mm, 270.6 mm, 283.6 mm, 297.6 mm, 312 mm'. The material shows a critical crack-opening displacement at relevant service temperatures of between 0.21 and 0.70 mm, and the stress in the deck should never exceed half the yield strength which is 240 MN/m², even under extreme sea conditions. Should the crack be repaired now, or can it safely wait for another four weeks when the ship will be in a port where better repair facilities are available?

Discussion

First establish the critical crack length under extreme conditions of highest stress, and minimum toughness, from *Figure 5.17*, plane stress conditions will be relevant. Hence

$$\delta = \frac{8}{\pi} \frac{240}{200 \times 10^3} \, a \ln \sec \frac{\pi}{2} \frac{1}{2} = 0.21 \text{ mm}$$

and

$$2a_{\text{CRIT}} = 397 \text{ mm}$$

The propagation probably follows a law like

$$da/dN = CK^n = C(\sigma\pi^{\frac{1}{2}})^n a^{n/2}$$

For a consistent pattern of stress cycles averaging out over each week, this could be restated as

$$\frac{da}{1 \text{ week}} = Ba^{n/2}$$

or

$$\log \frac{da}{1 \text{ week}} = \log B + \frac{n}{2} \log a$$

By plotting the logged information in the form of the weekly increase of crack length against the current crack length, it can be established that $n = 3$ and $B = 4.5 \times 10^{-3}$ give a reasonable fit to the data. Assuming that this trend can be extrapolated, and wishing to know the length of time required for the crack to reach 397 mm, the above equation can be inverted and integrated to give:

$$\frac{\text{weeks}}{da} = \frac{1}{Ba^{1.5}}$$

$$\text{No. of weeks} = \int_{312/2}^{397/2} \frac{1}{Ba^{1.5}} \, da$$

$$= \frac{10^3}{4.5} \times 2 \left[\frac{1}{a^{1/2}} \right]_{156}^{198}$$

$$= 4.0 \text{ weeks.} \qquad \text{Enough said!}$$

(N.B. Often the expression for K_I *cannot* be integrated analytically, and numerical or graphical methods must be used).

In the above example, it might be worth considering the effect of drilling holes at the crack ends. The strain range could be estimated from the 'blunt crack tip' stress intensity equations[17] or Inglis elliptical crack solution (reference 1, Chapter 5), and the problem could then be treated as a case of high-strain fatigue via the Manson/Coffin prediction.

6.4 Effect of other variables on fatigue life

6.4.1 Mean stress

The stress pattern on a fatigue-loaded component is usually separated for the purpose of analysis into a mean or steady stress level and an alternating stress range (see *Figure 6.5a*). For a given stress range, variation of the mean stress level has little effect on the *initiation* of cracks but can be important for *propagation*. A high tensile mean

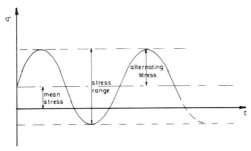

Figure 6.5a

stress (which implies a high maximum stress intensity in the cycle), tends to increase the propagation rate as already discussed and shortens the total life by bringing forward the intervention of fracture instability. It also reduces the effective threshold value of K (see *Figure 6.5b*). If the mean stress is compressive, on the other hand, propagation rates will be reduced—eventually to zero when the point is reached where the crack faces are pressed together throughout the loading cycle and it ceases to have significance as a crack.

Determination of the point where crack closure begins to reduce propagation rates is not particularly easy, as it is not safe to assume that closure occurs when the applied stress or stress intensity goes negative. As a result there is no clear convention even for presenting the results of fatigue crack propagation tests of the kind shown in *Figure 6.5b*. In some cases the full range of alternating stress intensity is quoted, accompanied by a *negative* value for the stress ratio R

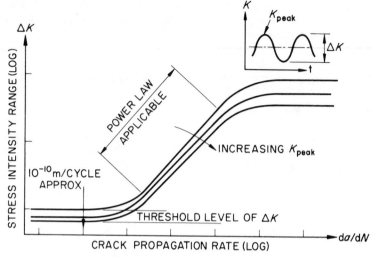

Figure 6.5b   Typical effect of tensile mean stress

$(\sigma_{max}/\sigma_{min})$ and in others the nominally negative part of the alternating intensity range is excluded.

In 'as-welded' structures, tensile residual stresses provide a significant source of harmful mean stress, and therefore it is commonly found that a component which has been repaired by welding fatigues all the sooner when it is put back into service. Conversely, stress-relieving treatments or other procedures such as spot-heating or peening, which actually introduce compressive residual stresses, can be beneficial.

6.4.2 Effect of multiaxial stress

Again it is important to distinguish the initiation and propagation phases of the fatigue process and neglect of this distinction has led to controversy and confusion concerning the appropriate characterising parameter in a multiaxial loading case. However, there is considerable evidence to show that for the initiation stage, multiaxial and uniaxial fatigue results can be brought into line if 'equivalent' stress or strain range (see p. 178) is used (Pascoe[18]). This should not be too surprising, as initiation is a shear-stress-controlled phenomenon and equivalent stress can be thought of as a 'root-mean-square' or 'average' shear stress.

However, when a crack is present and propagating, any

multiaxiality in the loading is overshadowed by the stress concentration at the crack tip, so that only the components of stress contributing to tensile stress normal to the crack plane are significant. (In most cases this simply amounts to the maximum tensile principal stress.) Of course stresses other than normal to the crack plane will have an effect on the plastic zone size and shape and to that extent there may well be a secondary influence on propagation rate, but at the present time the evidence is that the effects are marginal in most practical situations.

6.4.3 Straining rate (frequency effect)

The various phases of the fatigue process are controlled on the microscopic scale by the movement of dislocations and it has already been pointed out in Chapter 5 that dislocation movement is a time-dependent process, particularly in body-centred cubic materials. Thus it should not be surprising that fatigue damage and crack initiation should depend on the time allowed for plastic flow to take place, particularly at the high-strain end of the S/N curve. It is unfortunate that the effect is in this sense as laboratory tests are often conducted at higher than normal testing rates in order to complete within a practicable time and as a result falsely optimistic results may be obtained. The earlier warning concerning the effect of strain rate on the Manson correlation will be recalled and therefore one is forced to recommend that fatigue test results in the initiation phase especially, can only be considered valid for strain rates of the same order as that applied in the tests.

In the propagation phase, strain rate seems to be less critical, at least in the early stages when the plastic zone is small and plastic straining at the crack tip is limited (see Rolfe and Barsom[19]). In the later stages where ΔK is large, strain rate is again important as the plastic zone is large and rapid loading may influence the tendency to propagate by bursts of cleavage.

It is more realistic to think in terms of 'strain rate' as the controlling parameter rather than frequency of testing, as in the first place there may be a dwell period within the cycle, and also loadings of identical frequency with variable amplitude will produce different strain rates.

6.4.4 Temperature

Dislocation movement is encouraged also by thermal energy, so that a hotter environment has a similar effect to a slower strain rate. (This correspondence between temperature and strain rate has already been discussed in relation to fast fracture toughness.) There is one important exception to this trend in the case of carbon steels which strain-age strongly (Forrest[20]). Their fatigue strengths improve temporarily in the 200–300 °C range. Of course, if the temperature should be high enough to allow significant creep deformation, the problem becomes even more complex, as creep and fatigue interact strongly, especially in the high-strain regime.

In welded structures which are subject to thermal fluctuations, mechanical stress variations may occur due to incompatibility of thermal expansions. Thermal fatigue from this cause raises problems in furnace structures, piping systems and boiler parts, for example. Failure will often occur at the stress concentration points provided by weld shape and material discontinuities. The accumulation of cycles due to thermal variations is usually less rapid than for mechanical cycling in practical situations, hence thermal fatigue usually occurs at the high-strain end of the fatigue spectrum.

6.4.5 Environment: corrosion and associated effects

The atmosphere surrounding the welded component and perhaps penetrating cracks, has a strong effect on fatigue behaviour. Tests carried out in a vacuum give the longest life, which may be of some comfort to travellers in welded space vehicles. Dry gases, moisture, plain water and corrosive fluids have increasingly severe effects.

In the first place, corrosive media are likely to form pits in the surface which provide initiation sites, and weld toes are particularly vulnerable due to the combined effects of dissimilar metallurgy, notching and residual stress. When a crack is formed, the propagation rate tends to be much higher, especially in metals where hydrogen embrittlement is possible and in high-yield-strength alloys. Interaction with strain rate and test duration/frequency is also important, as corrosion and embrittlement processes are time dependent. Again there is a dilemma for the experimenter, as rapid accelerated tests will be optimistic in relation to real situations. Wei[21] provides a resumé of some of the effects in basic alloys.

Recent interest in the performance of welded steel structures offshore has led to the development of extensive research programmes into the fatigue response of materials and joints subject to salt water attack (reference 22). The effects of salt water are exceedingly complex and depend on the interaction of factors such as test duration, mean stress, impressed currents (such as employed in corrosion protection systems) and on the shape and depth of cracks.

Crack propagation rates are typically increased by factors of two to three times, although under adverse conditions the rates can go up to six times in the very low and very high propagation rate regimes.

It does however follow that any measure which is successful in excluding an aggressive environment from the joint or from a growing crack should be beneficial. Improvements have been claimed for plastic coatings and in the case of road decks, the asphalt covering is supposed to help matters. The problem as always with corrosion protection is to find a covering which is guaranteed to remain intact when the underlying metal is cyclically strained.

6.4.6 Variable amplitude loading

In most fatigue and crack propagation tests carried out in the laboratory, a constant amplitude and fixed mean stress loading programme is used. However, in real life the loading pattern is rarely so simple. In rotating machinery or where the loading is caused by vibration, the amplitude will tend to change with time (see *Figure 6.6a*) and in cranes, bridges, vehicles and handling machinery, both

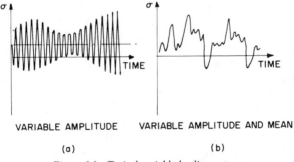

VARIABLE AMPLITUDE   VARIABLE AMPLITUDE AND MEAN

(a)   (b)

Figure 6.6   Typical variable loading patterns

the mean stress and the amplitude may vary randomly over a given period of use. Assuming that the pattern of loading is known or can be confidently predicted, there is still a problem in predicting the fatigue behaviour under variable amplitude loading in relation to constant amplitude behaviour, as it is not usually practicable to reproduce all possible variations of variable amplitude loading in a routine laboratory test.

The simplest approach to this difficulty is to assume that the concept of 'linear damage' is valid—that is that stress cycles of small amplitude make a small contribution to fatigue damage at a given point and large cycles make a large contribution, until enough damage has been caused at the point to lead to failure. This philosophy is expressed in Miner's rule which predicts failure in terms of n_i the number of cycles of a given magnitude and type actually occurring during a given period and N_i the number of cycles of that type required to produce failure in a constant amplitude test. Failure is assumed to occur for k different types of loading when the summation $\sum_{i=1}^{k} (n_i/N_i) = 1$. Of course, one must take care to ensure that the two sets of cycles are truly comparable in terms of mean stress, strain rate, etc., and neglect of this condition has contributed to experimental findings where failure occurred at $\sum (n/N)$ varying from 0.1 to 10. However there is a reasonable amount of empirical support for Miner's rule in practical situations and the very simplicity of the concept is appealing. Two assumptions implicit in the rule should be noted, however. Firstly, it can be seen that the order of application of loads does not affect the summation and, secondly, cycles below the fatigue limit (assuming there is one) are held to make no contribution to damage, as $N_i = \infty$.

Alternatively, if it is known or is assumed that the fatigue life is consumed exclusively by incremental fatigue crack propagation, variable amplitude situations can be assessed by piecewise integration of a suitable propagation law after the manner described earlier. One slight difficulty inherent in this treatment is that propagation laws are most often derived from data obtained from constant amplitude tests and in practice fatigue crack propagation depends on the sequence of loading. In particular, the inclusion of infrequent higher-than-normal loads in the spectrum can have the effect of delaying subsequent propagation relative to the relevant constant amplitude rate. The reasons for these so-called 'overload delays' are complex and not fully understood but much of the effect is due to the generation of an extra large plastic zone and a residual stress system which slows progress. Mechanical models for delay effects are currently being formulated and in time may find application, but for the present it may be noted that it will normally be conservative to ignore this effect in an assessment.

As it happens, although Miner's rule was advanced initially as a logical hypothesis and later supported empirically by testing, it turns out that the rule can be derived from a crack propagation model, provided that the initial and final crack integration limits are arbitrarily fixed (Maddox[23]). In practice, the final crack size can be varied substantially in accordance with appropriate failure criteria without affecting this conclusion significantly (Gray[24]). Thus the 'pure' fracture mechanics approach to variable loading is not necessarily much different from the empirical Miner's rule approach unless refinements which allow for sequence of loading and other specifics of the real situation are incorporated in the fracture mechanics analysis. Miner's rule is recommended in several British Standards for crane structures and bridges and has been incorporated in the DNV rules for offshore units[25].

6.4.7 Scale effects

Fracture mechanics contains an extremely important engineering message on the effects of increasing the dimensional scale of a component or structure; namely, that the crack propagation response of dimensionally similar size-scaled structures will not be identical. Even if triaxial transition effects due to variation in thickness are set aside, the basic crack-tip stress intensity will increase in proportion to the square-root of the crack size as all the dimensions are scaled up. Thus a crack which penetrates to 10% of the thickness is a more serious feature in a 200 mm thick structure than in a 20 mm thick scaled-down version, as the crack propagation tendency is at least $\sqrt{10}$ times greater. Most design codes which are based on a stress characterisation fail to recognise this trend and therefore sound no warning with respect to extrapolation of structural dimensions.

In the case of welded structures especially, development of an existing design by increasing the scale is a procedure which should be carried out with extreme caution as welding introduces further traps. Clearly a larger volume of weld brings a greater statistical risk of more significant crack-like defects but also the greater restraint and heat sink associated with thicker and stiffer sections often promotes higher levels of thermal and residual stress which in turn increases the likelihood of fabrication cracks and propagation in service. The effect of scale is especially noticeable with respect to fatigue crack propagation from geometrical stress concentrations. For example, if two geometrically similar components containing the same shape of stress concentration are compared as in *Figure 6.7*, a given *absolute* size of crack in the larger component experiences a higher level of crack-tip stress intensity and will propagate faster. Moreover, in the smaller structure, the crack may well stop propagating because it has run out of the high-stress region and is still below the threshold value for continued propagation as a stage II crack, whereas in the larger structure, the threshold condition is more easily reached. One implication of this side effect is that the fatigue limit strength of a larger-scale component will tend to be lower. Thus for example, the

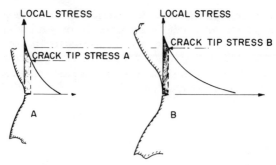

Figure 6.7 Geometrically similar structures with the same absolute size of crack

results of an S/N type fatigue test on a small-scale fillet welded detail are likely to be optimistic with respect to a larger-scale version of the same detail.

6.5 Practical approaches to fatigue-safe design

The previous sections outlined the important phenomena and trends revealed by fatigue research, and this information forms the background to the methods typically used to assess practical problems. In the end, designing against fatigue reduces to comparing the proposed design with known test cases and data, and it is recommended that the reader build up his own library or reference system for test information which is relevant to his own field. References 26 and 27 are good sources for data relating to welded structures.

There are usually three areas to be explored in a fatigue assessment: determination or prediction of the pattern of loading to be applied to the structure; stress analysis to relate these loads to stresses in the fatigue-prone locations; and comparison of the results with relevant test data. Just how these problems are tackled in particular industries depends very much on the background experience and volume of pertinent data previously generated. In the following sections a number of typical procedures and approaches are described and the chapter is completed by examining specific cases.

6.5.1 'Smooth' or 'defined stress concentration' details

In situations where a defined geometric stress concentration (see p. 201) is unavoidably included in a design, the fatigue strength/life of the component at the detail can be determined directly by comparison with fatigue tests on plain, smooth fatigue-test specimens of the relevant material. This circumstance most often arises in *machined* components rather than in welded structures, e.g. turned shafts, threaded components, and adjacent to drilled holes, but many welded structures are also machined to a smooth finish and in any case the maximum stress concentration effect may not be provided by the weld shape in a given location. For example, the maximum stress in a welded nozzle connection to a spherical pressure vessel often occurs not at the weld, but at the inside surface in a circumferential direction to the nozzle, at an 'as-forged' or machined point on the material.

Such cases can be treated simply by comparing the peak stress or strain range, given by the nominal stress/strain range × the elastic SCF, with a basic fatigue curve derived from the empirical correlations of Manson and others or from more specific data. Such an approach is adopted in BS 3915 and ASME section III for design of pressure vessel parts subject to high-strain/low-cycle fatigue. Material data relevant to the high-cycle regime and giving information on fatigue limits are given in reference 28. These data are mostly applied to fully machined components.

Estimating the SCF can often be quite difficult, despite the wealth of information available in the literature, especially where the characteristic notch radius is very small. It is worth remembering in this connection that sharp-crack stress-intensity solutions can be derived from the limiting case of radiused notch solutions as the radius tends to zero. The procedure can, if required, be reversed so that the SCF for a 'radiused crack', where the tip radius is small compared with the characteristic notch length, can be found from

$$\sigma_{\text{max}} = 2K_{\text{I}}/(\pi r_0)^{1/2}$$

where r_0 is the notch radius.

Notch sensitivity

If the stress *gradient* at the notch detail is very steep, it is sometimes pessimistic to assume that the full value of the SCF is felt by the component in terms of reduction in fatigue strength. (A steep stress gradient is normally associated with a small notch radius.) The effect can be satisfactorily explained in terms of fracture mechanics. Although the SCF and the resulting strain range at the notch surface may be high enough to initiate a crack, it propagates into a sharply reducing stress field so that the crack-tip stress intensity is insufficient to drive it forward quickly. At low stress levels an apparent fatigue limit may be found because the growing crack runs out of stress before it reaches the threshold level for continued propagation.

Clearly it is conservative to ignore this effect but in the past the phenomenon has been treated where thought necessary by assigning an empirical 'notch sensitivity factor' which varies between zero and unity (when the full SCF is considered to be effective). Alternatively data may be presented in terms of a 'fatigue strength reduction factor' given by notch sensitivity factor × SCF. Often this is designated K_f, with the implication that this is the magnitude of stress concentration *apparently* effective in fatigue, as distinct from the full geometric stress concentration factor K_t. The real problem in application is that these factors are empirical, material-dependent and become insignificant as the scale of the detail is increased (see Section 6.4.7). They are also of doubtful applicability in random amplitude loading. Reference 29 gives numerical values for different situations in machined parts.

As it is now accepted that the basic effect is due to the interaction of the stress gradient with a growing crack, it may be more logical if somewhat fussy, to treat the notch insensitivity effect in terms of a fracture mechanics model whereby an S/N curve is first used to determine the number of cycles to initiate a crack to a depth which allows continued growth above the threshold level (Smith[30]).

Surface finish is of course very important in relation to fatigue of 'smooth' details. Improvements in fatigue life which may be expected of materials with higher ultimate strength may not be fully realised unless scrupulous care is taken to grind smooth or to remove surface ripple and undercut effects in welds. In some cases the surface layer may have different properties due to under- or over-alloying and other factors caused by metallurgical history. Removal of this layer by light grinding can be beneficial, at least until corrosive or other environmental factors interfere with the surface.

6.5.2 Welded details

If the stress concentration effect in a given joint arises primarily from weld shape features or from the disturbed load transfer through the joint, the approach outlined in the previous section is simply not feasible because the numerical value of the elastic stress concentration cannot be estimated easily. Photoelastic and finite-element analyses of basic weld shapes have been carried out, and are documented in Gurney[26] for example, but weld profiles differ so much in practice from idealised geometries that it would be quite unrealistic to apply an exact numerical assessment to typical structural welded details.

The approach commonly adopted in such cases is to revert to the

S/N curve treatment, where data derived from the fatigue testing of real full-scale welded joints is applied. Such an approach has been adopted in BS 5400 for steel bridges, in BS 2573 for crane structures, and similar considerations form the basis of various design procedures for offshore structures, railway components and earth-moving machinery. The Welding Institute (formerly British Welding Research Association) at Abingdon pioneered the development and application of welded detail fatigue testing in the 1950s and 1960s and much of the data in use today comes from that source.

At first glance it might seem that each type of joint would require its own fatigue curve (or set of curves for different mean stress levels) but as different types of joint exhibit roughly similar fatigue strengths in the end despite differences in configuration and detailed shape variations, it is more convenient to classify different detail types into a relatively small number of strength groupings. The controlling parameter in applying these data is therefore the stress range—usually the tensile principal stress in a particular direction relative to the weld configuration—and although no stress concentration factor is explicitly used, the effects of overall geometry of the joint, of detailed fillet or butt weld contours and of typical cracks.

It is instructive to study the relative fatigue strengths of different details as presented in a typical design/assessment code, as the qualitative reasons for poor performance in certain kinds of joint can be better understood. *Figure 6.8* presents a paraphrase of the joint classifications contained in BS 5400, and *Figure 6.9* gives the associated design fatigue curves. Note that these curves do not represent raw data but were drawn approximately two standard deviations below the mean lines of typical constant amplitude fatigue test results. The full lines relate to experimentally based data and the broken lines represent extrapolations for long-life assessment. The tests were mostly carried out on medium-strength structural steels (400–600 MN/m^2 ultimate strength) and no increase is allowed for higher yield materials.

Class A represents the maximum possible fatigue strength attributable to the parent materials, that is in the machined condition assumed in Section 6.5.1. As this strength is theoretically unattainable in welded construction, no fatigue curve is given. The Class B details show that 'black' plate (as received from the steel mill) usually has scores and scale chips which produce a lowering of strength. Provided that there are no stress-concentrating features normal to the direction of loading, a condition which requires grinding of any weld overfill, fully penetrated continuous longitudinal welds can be included in this class.

286

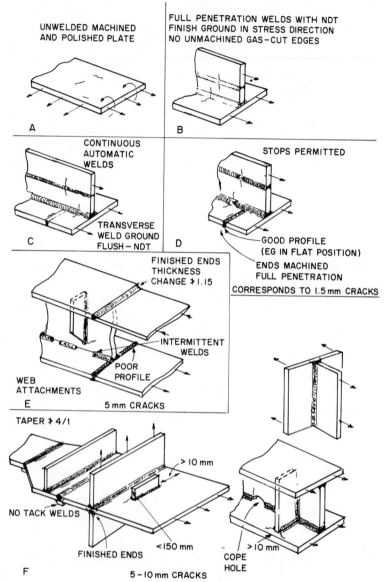

Figure 6.8   Fatigue classification of weld details

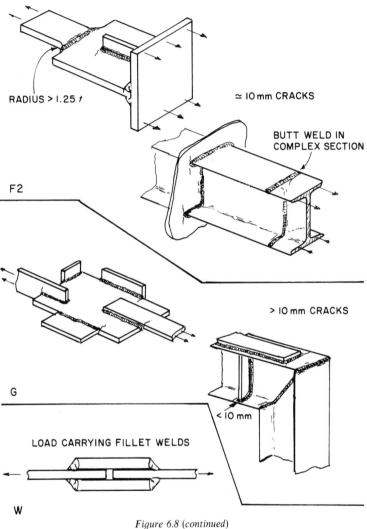

RADIUS > 1.25 *t*

≃ 10 mm CRACKS

BUTT WELD IN
COMPLEX SECTION

F2

> 10 mm CRACKS

G

< 10 mm

LOAD CARRYING FILLET WELDS

W

Figure 6.8 (continued)

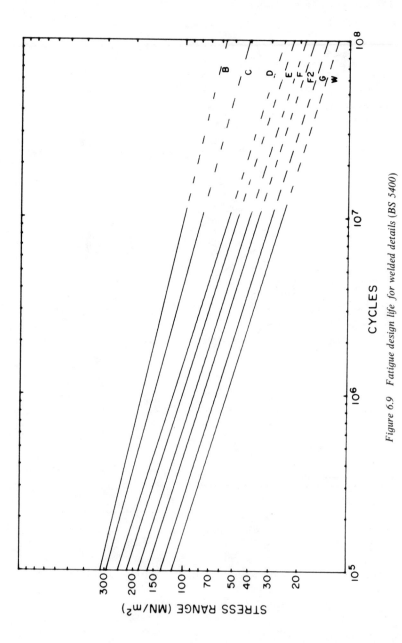

Figure 6.9   Fatigue design life for welded details (BS 5400)

Automatic welding procedures, carefully used, can provide greater freedom from weld ripple, stop/start notches and other defects. This potential is realised in class C which also shows that transverse butt welds may be included provided that the stress concentration normal to the stress direction caused by the overfill and any undercut is completely removed by grinding. Classes A, B and C must be essentially defect free, but a comparison of the quoted strength curves with Harrison[31] which explores the effect of slag inclusions shows that from class D downwards the fatigue strengths of each class are comparable with those of internally flawed welds where the pre-existing defects increase in size as shown. Alternatively, defects of the given size could be present without reducing the strength more than it is already. Maddox[32] has also observed that it is no coincidence that the curves for class D and poorer grades have a common slope; this conclusion can be drawn from a fracture mechanics model in which it is assumed that the total life is taken up with fatigue crack propagation from an existing defect. Thus only in classes A, B and C is a significant portion of the life taken up by initiation mechanisms.

The effect of overfill shape transverse to the stress direction is clearly seen in class D. If this form of stress concentration is combined with even a gradual change in thickness or width, further reductions are found as in class E which also highlights the deleterious effect of intermittent welding.

The picture becomes more complicated in considering classes F and F2 which were originally lumped together in an earlier standard (BS 153). Cracking may be expected at the ends of non-load-carrying fillet welded attachments and the tendency is greater as the length of attachment exceeds 150 mm. The difficulty of achieving full penetration butt welds in shaped sections such as I-beams, and the stress-concentrating effects of joint misalignment are also reflected in the F2 classification. Lap joints are downgraded to class G, a classification which, though the worst, still contains a wide variety of details in common use. This group also shows that attachments of any kind welded near the edges of a wide plate or flange are positively disastrous.

In many of the cases where fillet welds are required to transmit loads, it is clear that if the welds are insufficiently large, fatigue failure will develop from the root of the weld as in class W, rather than from the toe of the weld as shown in class F for example. Class W is also an addition to earlier standards, and gives an appropriate strength curve for weld metal, where the nominal principal stress *in the weld metal* is taken as the controlling parameter (see Section 4.2). Reference 33 gives good advice on the optimum size of fillet weld in such cases. Note that even this low classification does not include fillet welds subject to bending.

Again in earlier standards, adjustments to the data were recommended for different levels of mean stress—expressed in terms of the ratio $R = \sigma_{min}/\sigma_{max}$. As more data became available from tests on larger-scale structures, the feeling grew that the full effects of welding residual stress as a source of tensile mean stress had not been realised in the earlier tests and therefore the predicted beneficial effects of reducing mean stress might be illusory as far as real structures were concerned. Hence the present standards revert to the assumption that stress *range* is the important variable and only one set of curves relevant to an appropriately high mean stress is given. Some improvement might be expected for stress-relieved structures, particularly at negative values of R, but at present there appear to be no codes which allow for this.

The application of the approach embodied in the present BS 5400 has been very successful in improving the fatigue performance of welded structures, not least because it is educative in that it draws attention to detail configurations which have a poor performance. The information has also been freely applied to engineering design problems other than envisaged in the codes mentioned and the methodology outlined gives a sound framework for the application of other weld detail fatigue data, such as given in the compilations of reference 27.

Two cautionary notes should be sounded. Firstly, with regard to scale effect, the data which form the basis of the codes were obtained for the most part from tests of small size specimens (10–30 mm thickness, say) and as will be shown later in the case studies, these data may well be unconservative with respect to larger-scale dimensionally similar details. Secondly, very few tests have been carried out for practical reasons at low stress/long endurance, especially under random loading. The scatter and uncertainty in this region (greater than 10^7 cycles, say) can be quite large, as the situation tends to swither between propagation and non-propagation.

6.5.3 Analysis of load patterns

One of the major practical difficulties facing the designer relates to the problem of specifying or predicting the loads which will be applied to the structure. A few examples will make this point clear. In bridge structures, it is difficult enough to predict the likely growth of traffic in frequency and intensity over a 120 year period, but how far can nominal static axle loads and frequency of transit be related to stress amplitude and frequency in parts of the structure? In the case of ships and offshore structures, the mode of operation greatly influences the loads, apart from the variations caused by wind and wave actions for which reliable data may not be available. In the case of static load-carrying structures such as cranes and in rotating machinery such as fans, it may be feasible to define the specified load pattern with accuracy, but the possibility that parts of the structure may be excited to vibrate at higher frequencies in different modes needs to be considered carefully.

There are basically two approaches in common use for the prediction of loading. The first is to assume that the stress pattern is indeed directly related to the expected frequency of load application (as measured by traffic surveys in the case of bridges, or by lifting usage in the case of cranes, for example). The second method is to apply load or stress monitoring equipment to a typical structure of the type to be assessed and by collecting data over a representative period a picture is built up of realistic load spectra. The latter form of analysis has been in use for many years in aircraft engineering, but there is substantially less data available for common welded structures. A number of *ad hoc* tests are referenced by Gurney[26].

Taking the first 'design specification' approach, 'agreed' spectra for a variety of common structures can be found in several national standards and are worth consulting. For example, BS 5400 gives typical railway rolling stock loadings for railway bridges, and axle spectra appropriate to UK road bridges. BS 2573 gives specified loadings for various types of lifting devices and cranes. DNV rules[25] recommend methods for the calculation of statistical wave loads.

Application of strain measuring equipment and analysis of data is in itself a complex topic—a great deal of thought has to be given in the first instance to the placing of gauges and to the instrumentation and logging of data. When the results are eventually obtained, several interpretations can then be made in terms of stress amplitude, mean level and frequency of occurrence. Different analytical techniques for counting up the cycles of different amplitude have been proposed and are discussed at length by Gurney[26], but most logical from the point of view of picking out the maximum stress ranges is the 'rainflow' method (see later case study).

Two items of information need to be extracted through an analysis of loading, whether using an agreed arbitrary spectrum or from prototype data—the *frequency* of occurrence of a given stress range (mean stress level being a secondary parameter) and the *sequence* of occurrence. Again there are different ways of presenting the analysis, one important method being in terms of the *probability density* of a given stress range as in *Figure 6.10*. Probability density is the number of occurrences of a given stress range, expressed as a percentage of the total, and is divided by the stress-range interval (i.e. it is a finely divided histogram). Certain shapes of probability-density curve have

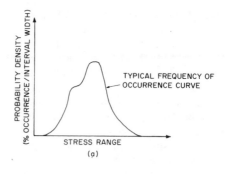

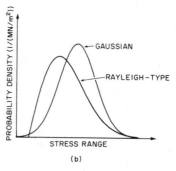

Figure 6.10

been found to be typical of certain structures and loadings. For example, if the loading is of a truly random nature with no cut-offs, the *Gaussian* distribution is appropriate (*Figure 6.10b*). Road surface undulations apparently constitute a good example of such a distribution. However, for various reasons the pattern of loading on structures is more often skewed towards the high-load end giving a *Rayleigh*-type distribution. This can happen in situations where loads are always above a given minimum but the maximum is either unlimited or relatively high. Frequently the coupling between the random phenomenon (ocean wave pattern or road surface) and the structure is such that a symmetrical input produces a skewed load pattern on the structure. Rayleigh spectra have been measured in short-span road bridges, and in offshore structures, wave-induced loads are often characterised by non-stationary skewed distributions where the mean level moves up as the weather changes from calm to storm.

However, it should be remembered that the probability-density curve does not itself give an immediate indication of fatigue damage;

as damage is roughly proportional to stress-range cubed, a smaller number of high-stress cycles may be relatively damaging.

In the case of load *sequence*, there appears to be no accepted way of characterising data. However, a few examples of typical situations may be considered, remembering that the important effect is the influence of overloads on propagation delays. Firstly, in cases where the amplitude varies randomly but *gradually*, e.g. as in *Figure 6.6a*, overload delays will be minimised and with the exception of threshold problems which will be discussed later, the linear superposition of constant amplitude data should be valid (Miner rules!). This occurs to a large extent in vibration-induced loading and to a lesser degree in weather-related phenomena. In traffic-induced patterns on the other hand, the sequence of individual loads must by definition be random and could only be assessed according to the laws of chance. This picture would also apply to multiduty lifting devices. One other case of interest is represented by the 'periodic overload' mode of operation. Many structures, pressure vessels being a good example, are subjected to an overpressure test at infrequent statutory intervals. These overloads are often too rare to contribute to damage accumulation, but should have the effect of delaying propagation.

Another important consideration in relation to random loading of welded details is the question of thresholds and fatigue limits. In the case of constant amplitude loading, the apparent fatigue limit stress is determined by the typical starter crack size and the threshold value of stress intensity. In random amplitude loading, stresses above the *initial* fatigue limit will extend the crack, so that more and more of the low-stress cycles will be effectively above the fatigue limit. In a recent random fatigue test series reported by the Transport and Road Research Laboratory, it was shown that 91% of the damage occurring in long endurance occurred *below the fatigue limit* (Tilly and Nunn[34]).

In such situations it is recommended either that the design fatigue curves should be extrapolated in a straight line or at some arbitrary angle below the fatigue limit (BS 5400).

6.6 Case studies

6.6.1 Fracture mechanics simulation of weld fatigue

There is considerable interest in simulating weld fatigue through a theoretical model of crack propagation, as it allows the effects of scale, material and other factors to be assessed in a more economical way than by full-scale fatigue testing. *Figure 6.11* outlines the basic approach. The initial problem specification is fed into a calculation of

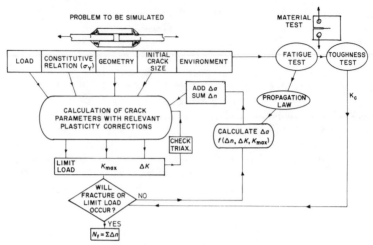

Figure 6.11   Simulation of weld fatigue

appropriate crack characterisation parameters, which uses LEFM or some appropriate plasticity approach such as the Dugdale model equations in Section 5.1.5. After checking that failure will not occur for the initial conditions, the increment of crack increase which will occur for a given number of cycles is calculated from a knowledge of the propagation law and added to the initial crack size to set off the calculation once more. Gray[35] gives full details of computer routines which can be used to facilitate calculation.

Figure 6.12 shows the results of a simulation of load-carrying T-butt welds compared against *S/N* data from Macfarlane and Harrison[36]. The crack model in this case was simply an internal crack of length equal to the specimen thickness, placed centrally in a square plate of total width equal to the crack length plus two weld throat thicknesses. This treatment ignores the stress concentration due to the angled weld configuration. Despite the simplicity of treatment, the prediction is good and a BS 5400 class W curve is shown for comparison.

Figure 6.13 attempts to predict the fatigue lives of transverse butt welds with small central slag inclusions (class D). Here the prediction is a little conservative, probably because the inclusions are not as sharp as real cracks and the mechanical or thermal stress-relieving treatments have benefitted the specimens. Note that the small difference in theoretical life attributable to yield strength is related to the use of a plasticity correction in the model.

Figure 6.14 simulates toe failures in fillet welds. The initial crack

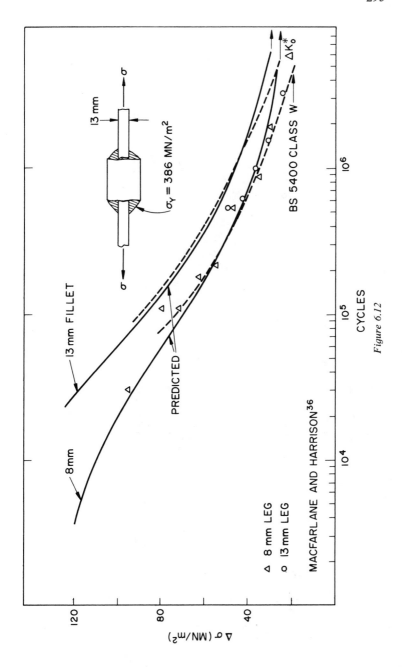

CYCLES

Figure 6.12

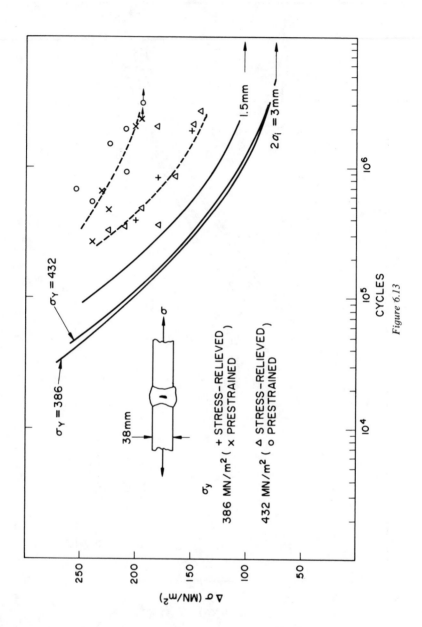

Figure 6.13

297

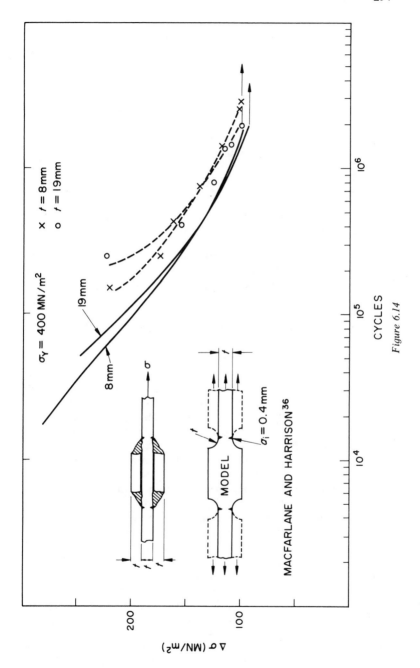

Figure 6.14

size has been set arbitrarily at 0.4 mm, in line with common observations (see Signes *et al.*[6]) and an allowance for the geometric stress concentration at the fillet weld toe has been made by using the solution for a crack growing from a circular hole.

Figures 6.15 and *6.16* demonstrate the effects of scale and geometry on life. Notice that the fatigue life *and the failure limit* reduce with increasing scale when geometrically similar details are tested. In these configurations and strengths, the end-point of the fatigue life is usually formed by net-section ductile failure (i.e. limit load); but in large thickness, the stress intensity may be high enough to cause plane-strain fracture.

6.6.2 Assessment of welded details in a mechanical handling structure

The case study described arose from an investigation following the failure in service of mechanical handling equipment. Although the failure was precipitated by the fracture of a particular welded bracket, the enquiry was extended to cover other welded details which were thought to be at risk. Various steps in the sequence of investigation are described to show how the information in Sections 6.5.2 and 6.5.3 can be brought to bear on a typical problem.

Firstly, the opportunity was taken to examine as many examples of the equipment as possible, operating in different locations. Particular attention was paid to potential crack sites as highlighted by the BS 5400 classifications. Cracks and indications of weld repairs were found, and a number of design detail variations were identified in equipment of different ages. There were several indications that the sense of the dynamic loads on the structure was not always similar to the static load pattern—for example, cracks were found in regions which should have been subject to compressive loading in the static case but were clearly in tension for typical operations. The machines were closely observed in normal use and information was gathered from operators on typical usage and experience.

In the light of these examinations, it was considered essential to carry out an experimental stress analysis of a representative machine under normal working conditions, including dynamic loading. A preliminary theoretical analysis was carried out to provide a basis for the strain-gauge layout plan and to give information for static checking of the strain gauges. *Figure 6.17* shows the main welded details examined and indicates the positions of the key strain gauges. The main form of construction consisted of rectangular rolled-hollow-section members, stiffened by the addition of cover plates and

299

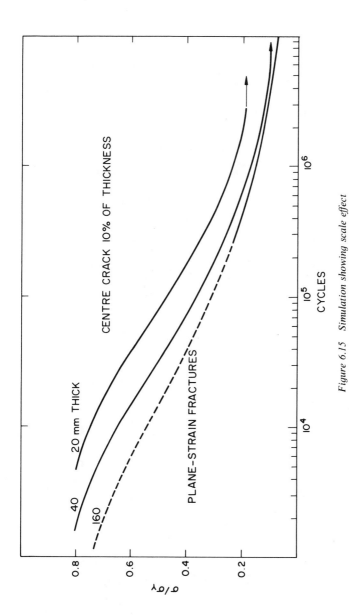

Figure 6.15 Simulation showing scale effect

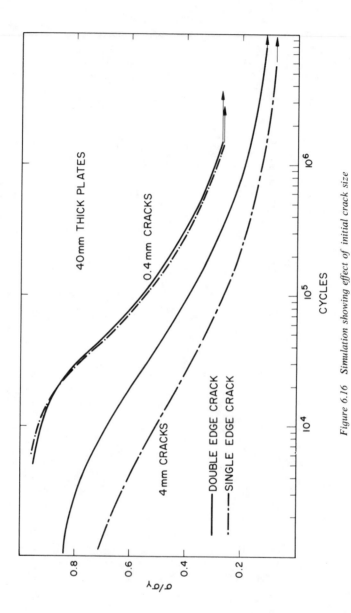

Figure 6.16   Simulation showing effect of initial crack size

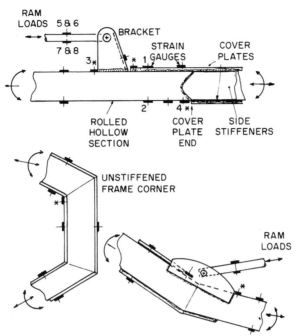

Figure 6.17 Welded details on a mechanical handling structure

side plates, with fabricated brackets attached to react hydraulic ram loads.

The strain gauges were placed to give three different kinds of indication: first, the loads applied by rams; secondly, the 'nominal' bending and direct stresses in the arms at locations sufficiently removed from the stress-concentrated areas to correspond to the principal stress locations which provide the reference stress levels for the BS 5400 classifications (see *Figure 6.8*); and thirdly, some gauges (marked *) were placed in conjectured peak stress locations— especially at points where cracks or repairs had been found. (Note that accurate estimation of SCFs demands a much finer measurement mesh than used here.) The nominal stress gauges also allowed parts of the structure to be used as 'load cells' in that the magnitudes and lines of action of forces could be deduced from the measured strains and the section properties. Notice also that rosettes were not used in this instance, as the principal stress directions could be confidently predicted (see Section 4.1.8).

The structure was then subjected to various loading sequences which served to check the basis of theoretical stress analyses and to determine a base-line for dynamic testing. It was then put through a series of normal operations with a calibrated load and the dynamic strain patterns recorded. A typical portion of dynamic record is shown in *Figure 6.18*.

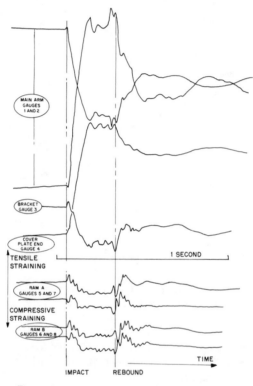

Figure 6.18   Portion of strain gauge dynamic record

The raw data from the tests were then processed to build up a representative stress/cycles profile for each detail in the structure. For example, the nominal principal stress pattern at the end of the cover plane in *Figure 6.17* was reconstructed as in *Figure 6.19*. A 'rainflow' analysis of the pattern then produced the following breakdown of cycles for a typical operation.

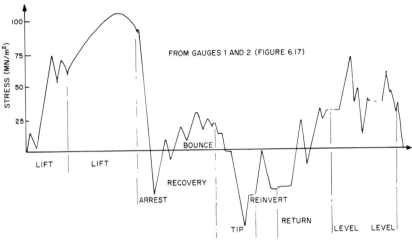

Figure 6.19   Typical stress pattern

$\Delta\sigma$ (MN/m²)	σ_{mean} (MN/m²)	*Cause of cycle*	*N (class G)* mean line
166	12	lift → tip	1.24×10^5
73	37	level	1.47×10^6
55	39	level	3.43×10^6
55	1	arrest → recovery	3.43×10^6
36	6	return	1.22×10^7
31	−15	re-invert	1.9×10^7
22	66	lift	5.35×10^7
15	1	arrest/bounce	1.69×10^8
		$\sum 1/N$	9.5×10^{-6}

Hence the number of typical operations which will produce failure at this detail according to Miner's rule will be $1/9.5 \times 10^{-6}$ which is 105 000 operations. The allowable design life for this detail (the design fatigue curves being two standard deviations below the mean data) is 45 000 operations, which for the particular structure considered amounted to $2\frac{1}{2}$ to 4 years service, depending on usage.

6.6.3 Design/repair assessment

Example 6.1

Three alternative designs are proposed for stainless steel heat exchanger tube-to-tubeplate welds (see *Figure 6.20*). The most

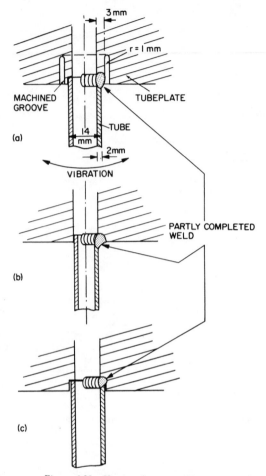

Figure 6.20　Heat exchanger welds

significant loading in such cases is alternating bending, due to vibration. Estimate the lives of the three details for an applied bending moment of ± 40 N m which produces a nominal bending stress of ± 6.25 MN/m² in the tube. The ultimate tensile strength of the austenitic steel is 500 MN/m².

Discussion

The tubular configuration and the material used here will invalidate the strict application of BS 5400, but such data nevertheless provide a guide. The weld quality is probably better than normally obtained for structural manual-metal-arc welds.

The weld in detail (a) probably corresponds best to class E in that although it is a transverse butt weld with a good profile, there is a change of thickness which will tend to increase the stress concentration effect. From BS 5400, a design life of about 5.3×10^5 cycles might be expected. Gurney and Maddox[37], who deal with the fatigue of tubular configurations, suggest an experimental life of 8×10^5 cycles. The radius at the top of the machined groove is also at risk in this design. Treating this for design purposes as a shouldered tube with a large change in external diameter, reference 29 could be used to estimate the SCF at about 2. Taking into account the difference in thickness between the tube and the stub, the peak elastic stress range at the sharp corner would be about 168 MN/m² ($2 \times \frac{2}{3} \times 125$), which is close to the estimated fatigue limit for this material. Note however that the 3 mm thick tube stub is being stressed normal to the plate rolling direction, and there is a very real risk of premature fatigue failure through a laminar weakness in this design.

Design (b) probably introduces a more severe stress concentration in the weld joint than design (a), as the two joined components are poorly aligned longitudinally. Classification into classes F or F2 would seem more appropriate in this case, giving a design life of 2.2×10^5 to 6.3×10^5 cycles. Notice that in both assessments no allowance has been made for the low level of mean stress in this example (σ_{mean} is zero), and it seems likely that in this instance there should be some conservatism from this source.

Detail (c) contains a built-in crack, albeit roughly parallel to the stress direction. By analogy with an 'infinitely-thick permanent backing strap' it might be regarded as belonging to class F.

Example 6.2

A wide flat steel tie bar (see *Figure 6.21*) is loaded 100 times per day from zero to 180 MN/m². After 8 years service it is found that a fatigue crack has developed from the sharp machined corner as shown. Surface fillet welds are applied to effect a temporary repair. How long might they be expected to last? Would full penetration welds be better?

Discussion

The detail in *Figure 6.21a* may be treated as a 'smooth' stress concentration and the SCF estimated from reference 9 of Chapter 4 as 2.3 ($D/d = 2$; $r/d = 0.1$). Hence the strain range is calculated to be

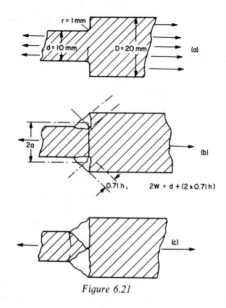

Figure 6.21

$\Delta\varepsilon = 2.3 \times 180/200 \times 10^3 = 0.204\%$. *Figure 6.2* predicts a life of 2.5×10^5 cycles which is about 7 years life.

The surface fillet weld repair will probably fail from the weld root as shown in *Figure 6.21b*, as the nominal weld stress is greater than the tie bar stress. The reference stress for class W is given by $\Delta\sigma = 180 \times (10/2 \times 0.707 \times 5) = 255$ MN/m². The predicted life from **BS 5400** is 22.3×10^3 cycles or 223 days, and the allowable design life is 40% of this.

In the case of the full penetration repair shown in *Figure 6.21c*, the final failure site will transfer to the minimum cross section at the weld toe. This type of detail may be classified as type E for which a much improved life of 5.6×10^5 cycles is predicted (which is probably better than the original). Of course the *ends* of the welds in both (b) and (c) techniques must be *properly* finished and ground to ensure that no cracks start from these points.

REFERENCES

1. MANSON, S. S., 'Fatigue; A complex subject, some simple approximations', *J. Exp. Mechs* (1964)
2. MILLER, K. J., 'Cyclic behaviour of materials', *J. Strain Analysis*, **5**, No. 3 (1970)
3. LANDGRAF, R. W., *et al.*, 'Determination of the cyclic stress–strain curve', *70th Annual meeting of ASTM* (1967)

4. SPENCE, J., and CARLSON, W. B., 'A study of nozzles in pressure vessels under pressure fatigue loading', *Proc. Instn Mech. Engrs*, **182**, pt 1 (1967)
5. 'Nuclear Vessels', ASME Section III (1965)
6. SIGNES, F. G., *et al.*, 'Investigation of the factors affecting the fatigue strength of welded high strength steels', *British Welding J.*, **14** (No. 3), 108 (Mar. 1967)
7. LAIRD, C., 'The influence of metallurgical structure on the mechanisms of fatigue crack propagation', ASTM Special Tech. Publ. 415, 132
8. HARRISON, J. D., 'An analysis of non-propagating cracks on a fracture mechanics basis', *Metal Construction*, **2**(3), 93–8 (1970)
9. TOMKINS, B., SUMNER, G., and WAREING, J., 'The effect of material stress–strain properties on fatigue fracture', Paper 62, *2nd Int. Conf. on Structure*, Brighton, April, 1969, Chapman and Hall
10. GURNEY, T. R., *Fatigue of Welded Structures*, Cambridge (1968)
11. JACK, A. R., and PRICE, A. T., 'The use of crack initiation and growth data in the calculation of fatigue lives of specimens containing defects', *Metal Construction*, **3**, No. 11, 416–419 (Nov. 1971)
12. HARRISON, J. D., 'The analysis of fatigue test results for butt welds with lack of penetration defects using a fracture mechanics approach', BWRA report E/13/67
13. MADDOX, S. J., 'Fatigue crack propagation in weld metal and HAZ', *Metal Construction*, **2**, No. 7, 285 (July 1970)
14. MADDOX, S. J., 'Calculating the fatigue strength of a welded joint using fracture mechanics', *Metal Construction*, **2**, No. 8, 327 (Aug. 1970)
15. BROTHERS, A. J., and YUKAWA, S., 'Fatigue crack propagation in low alloy heat treated steels', ASME paper 66, Met 2
16. CLARK, W. G., 'Subcritical crack growth and its effect upon the fatigue characteristics of structural alloys', *Engng Fracture Mechanics*, **1**, 385–397 (1968)
17. CREAGER, M., and PARIS, P. C., 'Elastic field equations for blunt cracks', *Int. J. Fract. Mechs*, **3**, 247 (Dec. 1967)
18. PASCOE, K. J., 'Low cycle fatigue in relation to design', *Fracture*, Chapman and Hall (1969) (see discussion)
19. ROLFE, S. T., and BARSOM, J. M., *Fracture and Fatigue Control in Structures*, Prentice Hall (1977)
20. FORREST, P. G., *Fatigue of Metals*, Pergamon (1962)
21. WEI, R. P., 'Some aspects of environment enhanced fatigue crack growth', *Engng Frac. Mech.*, **1**, 633–51 (1970)
22. *European Offshore Steels Research Seminar*, Welding Institute (1980)
23. MADDOX, S. J., 'A fracture mechanics approach to service load fatigue in welded structures', *Welding Res. Int.*, **4** (2), 1–30 (1974)
24. GRAY, T. G. F., 'Comparison of crack propagation superposition model with Miner's Rule', *Research Report, University of Strathclyde* (1980)
25. 'Rules for the construction and classification of mobile offshore units', Det Norske Veritas (1975)
26. GURNEY, T. R., *Fatigue of Welded Structures*, 2nd edn, Cambridge University Press (1979)
27. *Fatigue Strength of Welds: Stress and Strength Sub-series*, Vol. 5, Engineering Sciences Data Unit
28. 'Fatigue strength of materials', Items 71027, 73005, 74016, 74027, Engineering Sciences Data Unit
29. 'Design against fatigue-basic design calculations', Item 75022, Engineering Sciences Data Unit
30. SMITH, R. A., 'A simplified method of predicting the rates of growth of cracks initiated at notches', *Fracture Mechanics in Engineering Practice*, Applied Science Publishers (1977)

31. HARRISON, J. D., 'Basis for a proposed acceptance standard for weld defects', *Metal Construction*, **4**, 7 (1972)
32. MADDOX, S. J., 'Fracture mechanics applied to fatigue in welded structures', *Proc. Conf. on Fatigue of Welded Structures*, Welding Institute (1971)
33. 'Fatigue strength of transverse fillet and cruciform butt welds in steels', Item 75016, Engineering Sciences Data Unit
34. TILLY, G. P., and NUNN, D. E., 'Variable amplitude fatigue in relation to highway bridges', *Proc. Inst. Mech. Engrs*, **194**, 17 (1980)
35. GRAY, T. G. F., 'A closed form approach to the assessment of practical crack propagation problems', *Fracture Mechanics and Engineering Practice*, Applied Science Publishers (1977)
36. MACFARLANE, D. S., and HARRISON, J. D., 'Some fatigue tests of load-carrying transverse fillet welds', *Br. Welding J.*, **12**, 613 (1965)
37. GURNEY, T. R., and MADDOX, S. J., 'Determination of fatigue design stresses for welded structures from an analysis of data', *Metal Construction*, **4**, 11 (1972)

Index